AF715146

The Earth Doesn't Move

Sandeep Chavan

Published by SANDEEP CHAVAN, 2025.

THE EARTH DOESN'T MOVE

First edition. July 16, 2025.

ISBN: 979-8231517015

Written by Sandeep Chavan.

Table of Contents

1: The Earth Doesn't Move – But You Do 1
2: Newton's Fallacy 6
3: The Inertial Frame Illusion 11
4: The Constant Arrow of Time 17
5: Force: A Symptom, Not a Cause 23
6: Mass and Matter 29
Interlude: The Stillness Beneath the Storm 36
7: Spin Deception 39
8: Orbit Fallacy 46
9: Gravity: From Attraction to Ripple Pressure 53
10: Why You Don't Feel the Spin 60
11: Foucault, Gyroscopes, and the Misread Proofs 66
12: Satellites and the Spaceframe Illusion 73
Interlude: When the Ground Refuses to Move 80
13: Redefining Inertia 83
14: Acceleration 90
15: Motion Redefined 97
16: Energy 104
17: Momentum 111
18: Light Doesn't Travel – It Appears 119
19: Time and Motion Are Not Fundamental 126
20: Electromagnetic Field 132
21: Field Healing 139
22: Consciousness 147
Prelude: When We Finally Looked Up 152
23: The Big Bang Isn't a Beginning 155
24: What Einstein Missed 159
25: Quantum Myths 162
26: The Observer Illusion 166

27: Stillness Is the Final Law 169
28: The Return to Common Sense – But Sharper 173
Epilogue: The Stillness We Forgot 177
Author Reflection 179
Appendix A: Glossary of Classical vs UED Terms 182
Appendix B: Newton's Own Doubts and Letters 187
Appendix C: Suggested Experiments 190
Appendix D: Quotes to Question Motion 195
Appendix E: References and Further Reading 199
Acknowledgments 203
About the Author 205

To all the scientists, philosophers, seers, and silent observers—from the ancient sages who mapped the stars without ever needing them to move,to the pioneers of reason who dared to ask, "What if everything we see is a reflection?"to Newton, who saw the apple fall,to Einstein, who bent time to explain it,and to every modern thinker—famous or forgotten—who tried to unify reality through logic, equations, or intuition.

This book is for those who questioned the obvious,who dared to doubt the dogma,who honored perception before theory,and who understood that science was never meant to end—it was meant to evolve.

May we all remain students of stillness. And may our ripples, however small,help resolve the great field of understanding.

— *Sandeep Chavan*

"The sky has never moved.The stars have never left.It was our belief that kept spinning."

— *Sandeep Chavan*

From Stillness, Through Ripple, Into Clarity

Disclaimer

This book is a work of independent thought, built on original research, reinterpretation of existing scientific models, and speculative reasoning grounded in the Universal Energy Dynamics (UED) framework. It challenges widely accepted scientific paradigms not to provoke, mislead, or deny existing evidence, but to offer a fresh lens through which reality may be re-examined.

While every effort has been made to present the ideas with logic, clarity, and intellectual honesty, readers are encouraged to approach the content with an open yet critical mind. The views expressed are those of the author and do not represent any academic, institutional, or governmental consensus.

This book is not intended to serve as a substitute for formal scientific education, nor should it be construed as rejecting the value of established scientific work. Rather, it is an invitation to explore the unresolved, question the inherited, and rethink the assumed—from a place of inquiry, not ideology.

The author accepts no responsibility for how the material is interpreted, debated, or applied. The journey of perception belongs to the reader.

Opening Note

The Earth Doesn't Move — But Your Mind Will.

For over 300 years, we've built towers of science on a foundation no one questioned: that the Earth is in motion, spinning at unimaginable speed, orbiting the Sun in a cosmic ballet of precision and predictability.

We applauded Newton for replacing Aristotle. We worshipped Einstein for curving space and time. And we accepted, without protest, the idea that motion defines everything — from falling apples to exploding galaxies.

But what if the foundational assumption was never true? What if motion isn't the cause of experience, but the effect of alignment in something far deeper — something still?

This book is not about flat Earth theories or anti-science rants. It is about reclaiming the right to question — to dismantle long-held assumptions and see the world with fresh, rational eyes. Not eyes clouded by authority, but sharpened by logic.

Here, you will encounter a new framework — Universal Energy Dynamics (UED) — that sees the universe not as a stage of moving objects, but as a structured field of stillness where ripples replace motion, and consequence replaces cause.

We will revisit every fundamental principle — inertia, force, gravity, time, light — and reveal what lies beneath their textbook definitions. Not to destroy science, but to resurrect truth. Not to belittle Newton, but to show that even he disguised Earth's stillness beneath a mask of mathematical convenience.

Prepare to unlearn. Prepare to be uncomfortable. Prepare to move — even as Earth stays still.

This book is not here to convince you. It's here to awaken you.

Let's begin.

— *Sandeep Chavan*

Independent Researcher, Ripple Realist

Prologue

When the Ground Beneath You Lies

They taught us that we're on a spinning rock, hurtling through space at 1,600 kilometers per hour — and yet, we don't feel a thing.

They said we orbit the Sun at 30 kilometers per second — and yet, sunrise feels like it waits for us.

They mapped the cosmos, curved space, and sliced time. And we — trusting, obedient — clapped.

We clapped for laws we never questioned. We memorized constants we never understood. We parroted proofs we never dared to test.

But when you sit in silence, when you breathe in the stillness of a windless night, when you feel the steadiness of Earth beneath your feet — what does your body tell you?

It tells you that the Earth doesn't move. And it never did.

Not because it's flat. Not because it's mystical. But because movement is a ripple illusion, a resolved consequence of deeper interactions. Stillness is the base layer of all things.

What you call motion is the echo. What you call force is the memory. What you call science is the interpretation — not the truth.

In this book, we will take you back to that truth. Not through rebellion — but through revelation. Not through fantasy — but through field logic.

We will challenge Newton, not to dismiss him, but to complete what he never dared to finish. We will question Einstein, not to mock him, but to anchor his genius in stillness.

This is not a theory. This is a reckoning.

Because one day, you must ask yourself: If the Earth is spinning, why do I feel so still? And if I am still...Then what, exactly, is moving?

Welcome to the untold story beneath every step you've ever taken. The Earth doesn't move.

But your perception — that's where the real motion begins.

Part I – The Foundations of Fallacy

1: The Earth Doesn't Move – But You Do

The introductory paradox and the trap of inherited belief

You wake up in the morning. The sun is rising. The trees are still. The cup of tea in your hand doesn't slide off the table. The birds fly in elegant arcs, the wind blows gently, and everything feels... grounded. Stable. Calm.

And yet, we're told you're on a spinning ball. Spinning at 1,674 kilometers per hour at the equator. Revolving around the Sun at nearly 108,000 kilometers per hour. Hurtling through the galaxy at over 828,000 kilometers per hour. All while the galaxy itself races through space at even more mind-bending speeds.

You're not just moving. You're a bullet on a bullet on a bullet. And somehow, your coffee stays in the cup.

You're expected to believe this. In fact, you were never given a choice.

This is the paradox. The Earth doesn't feel like it's moving — because maybe it isn't. And the only thing that has been moving, all along, is your belief.

Inherited Knowledge: The Trap We Call "Science"

Before we challenge any law, let's challenge ourselves.

Ask yourself honestly — *how do I know the Earth is moving?* Not what you memorized. Not what the textbook printed. Not what the teacher said.

What do you actually observe?

The answer is: *Nothing.* You don't observe Earth's rotation or revolution. You experience day and night. You see the Sun rise and set. But that could also be explained by a still Earth and moving Sun — as ancient cultures believed for thousands of years.

You didn't reject that view because you proved it wrong. You rejected it because someone told you it was wrong — and gave you a "better" story.

That story became belief. That belief became assumption. That assumption became law. And that law became "science."

But inherited science is not the same as true understanding. When knowledge becomes dogma, it loses its truth-seeking edge.

The Subtle Switch: When Belief Became Law

In Newton's time, people still debated whether Earth was moving. Newton didn't explicitly say *Earth is spinning* — instead, he quietly introduced the inertial frame of reference, where certain objects are assumed to be in a "non-accelerating" frame.

And what frame was more convenient to assume as inertial than Earth itself?

But this was a disguised assumption, not a proven fact. He built his entire physics model on it. And we — centuries later — continue to do the same.

We believe we're using logic and mathematics, but we're actually using logic on top of an unproven base.

In simple terms:

Newton built the house of modern physics using a foundation he declared stable — but never checked.

That's not a crime. That's how early models work. But refusing to question that base once new insights arise? That's how fallacies survive.

Let's Talk Stillness

What if the Earth actually is still?

Let's drop all preconceptions and follow observable logic:

- You feel no spin.
- You feel no orbital tug.
- You drop a ball — it falls vertically.
- You fly in a plane — you don't feel a 100,000 km/h background rush.
- You sleep peacefully while "moving" faster than any rocket ever built.

This isn't flat-Earth logic. This is pre-assumption observation — the kind true scientists begin with.

Stillness is the first thing you know. Stillness is what you're born into. Stillness is what your body trusts — what your senses confirm.

Motion, on the other hand, is told to you.

So ask: Why do we override stillness in favor of motion?

Enter the Ripple Reality

Now comes the shift — from what *appears* to what *is*.

In **Universal Energy Dynamics (UED)**, motion is not a primary cause — it's a ripple response in a structured vacuum field. Everything that seems to move is actually a consequence of ripple imbalance resolving back toward stillness.

So, when we say an object "moves," we're not watching it "go" anywhere. We're seeing a ripple participate in a field interaction that creates the *appearance* of motion.

The Earth? It doesn't move. It holds the field. It's the platform of ripple anchoring in our observable frame.

When you throw a ball, it's not the ball that moves through empty space. It's the ripple state changing through structured vacuum, as per your intent.

In this sense, the Earth doesn't rotate — it anchors ripple equilibrium. It doesn't revolve — it hosts the ripple interactions that create rhythmic perception (day, night, season, etc.).

The Real Movement: You

So if the Earth doesn't move... then what does?

You do.

Not your body — your beliefs, your frames, your assumptions.

Your thoughts move through time. Your ideas evolve through layers of illusion. You accept models, defend them, teach them, build machines on them. But you rarely stop to examine the floor beneath your mental feet.

In truth, it's *you* that's spinning — not the Earth. You're spinning in intellectual tradition. You're revolving around theories you've never personally tested. You're trapped in inherited belief disguised as truth.

And the worst part? You call this "education."

This Book Is Not Here to Convince You

Let's be clear. This book is not here to convert you into a still-Earth believer. It's here to remove your blindfold.

It's here to say: "You were never taught to think like this. But now you can." It's here to bring back scientific rebellion with clarity and conscience.

The Earth doesn't move. Not because it can't. But because it doesn't need to.

All that moves is field tension, ripple alignment, and your perception — dancing through structured stillness.

Final Thought for This Chapter

Newton gave us the gift of laws. But he also left us with a silent, disguised assumption — that Earth can be treated as stationary for convenience, but must be considered moving for cosmology.

We accepted the contradiction. And from that seed, a forest of confusion grew.

Now it's time to uproot that forest. Not with anger, not with rebellion — but with stillness. Because stillness is not laziness. It's clarity.

And once you see clearly...You'll never unsee the stillness beneath all movement.

2: Newton's Fallacy

Mistaking Consequence for Cause

Ripple-aligned inertia vs. object's motion

Every great thinker in history has built upon the ruins of the one before. Newton challenged Aristotle. Einstein challenged Newton. And now, we challenge them both — not to erase their brilliance, but to reveal what they couldn't yet see.

When Newton shattered Aristotle's idea that motion requires a continuous push, he brought something revolutionary to the table: Inertia — the idea that an object in motion stays in motion unless acted upon.

This seemed radical. Elegant. Empirically grounded. But it came with a silent cost — a new illusion wearing the robes of logic.

In trying to break free from Aristotle's fallacy, Newton created one of his own: He mistook the consequence of field alignment for an internal property of the object.

And thus was born what we now call: Newton's Fallacy.

What Is Newton's Fallacy?

In plain terms: Newton's Fallacy is assuming that motion is a property carried by the object, rather than a *ripple state aligned within a still universal field.* It mistakes the consequence of ripple equilibrium for the cause of continued motion.

When Newton saw an object glide across a surface or fall through the air, he defined its motion as a result of initial force, sustained by inertia, and modifiable only through interaction (friction, gravity, or applied force).

But in UED terms — that's a shallow resolution of a much deeper phenomenon.

Inertia Isn't What We Thought

In Newtonian physics, inertia is:

- A tendency to maintain current motion or rest.
- An intrinsic property of mass.
- The reason objects keep going unless stopped.

It's a law we've accepted, repeated, and built machines upon.

But think deeper:

- Where does this "tendency" come from?
- Why would an object resist a change?
- What mechanism *sustains* the motion in a vacuum?

Answer: It doesn't. The motion sustains itself only because the field it participates in remains undisturbed. The object doesn't "have" inertia. It is simply ripple-locked in equilibrium with the surrounding field.

Ripple-Aligned Stillness: A New Definition of Inertia

In the Universal Energy Dynamics (UED) framework, we redefine inertia:

Inertia is the ripple-locked equilibrium of an object in a structured still field. The object doesn't "possess" motion. It participates in a ripple pattern that continues until disturbed.

This means:

- Motion is not intrinsic. It is emergent.
- Nothing "wants" to stay in motion. It just doesn't meet resistance until a new ripple state interferes.
- Every observed velocity is a passive consequence, not an active trait.

A Simple Analogy: The Leaf on a Pond

Imagine a calm, still pond. You drop a leaf. If you create a gentle ripple beneath the leaf, it begins to drift — not because the leaf has motion, but because it is aligned with a ripple flow.

Now remove the ripple. The leaf stops. Not because something *acted* on it, but because nothing sustains its movement anymore.

This is ripple-aligned inertia. Not a force. Not a tendency. Just participation in a field state.

Newton's View vs UED: A Side-by-Side Thought Shift

Concept	Newtonian View	UED View
Inertia	Internal resistance or tendency	External ripple equilibrium
Motion	Object continues unless stopped	Object appears in motion while ripple alignment is stable
Force	External action changing motion	Ripple distortion triggering realignment
Mass	Source of inertia and resistance	Degree of ripple condensation within field

This reframing changes everything.

Why Newton Didn't See It

It's easy to criticize from the future. Newton didn't have the language of fields, let alone structured vacuum energy. His genius was in creating a framework that *worked* — for objects, machines, and early cosmology.

But there's a difference between what works and what's true. Classical mechanics works like a surface map. But UED shows us the ocean underneath.

Newton's equations are valid approximations within ripple-stable systems, but they fail to explain:

- Why motion continues *without a carrier*.
- Why "rest" is even possible in a rotating universe.
- Why energy appears conserved without an engine to sustain it.

Because these were never object phenomena. They were always field phenomena.

Let's Test the Assumption

Take a ball, roll it across a frozen lake. It continues moving for a long time. Newton says: inertia is keeping it going.

Now imagine the same lake, but in a high-ripple zone. Suddenly, wind gusts, vibrating ice, distant footfalls — all cause subtle field interference. The ball stops sooner.

Did inertia change? No. The field alignment changed.

The ball didn't "resist" motion. It simply lost participation in a ripple equilibrium that once sustained it.

This Is More Than Semantics

This isn't just a new way to talk about old things. It's a fundamental paradigm shift in how we understand reality.

Because once you accept that inertia is a ripple consequence, not an internal trait:

- Motion is no longer mysterious.
- Stillness becomes the natural state.
- All motion becomes traceable to field interactions.
- And most importantly, the Earth can be still — without contradiction.

Suddenly, Newton's mechanics become special cases, not universal truths.

Final Thought for This Chapter

We do not discard Newton. We honor him by finishing the work he unknowingly left open.

He broke the grip of Aristotle by stating:

"Things don't need a push to keep moving."

We now say:

Things don't move at all — unless ripples carry them. And once ripples stabilize, movement ends — not because of force, but because of resolved consequence.

Motion is not a truth. It is a signal. And when we finally learn to read the ripples instead of chasing the object, we'll understand: Inertia was never a mystery. We just mistook the echo for the cause.

3: The Inertial Frame Illusion

Why all frames assume a still Earth — without ever admitting it

In every classroom, the phrase is repeated like a sacred chant: "A body at rest or in uniform motion will remain so unless acted upon by a force — in an inertial frame of reference."

Simple, right? Clean, elegant, universal.

But ask a physics teacher — *What exactly is an inertial frame?* And you'll get a definition that quietly sidesteps the real question:

"An inertial frame is one in which Newton's laws hold true."

Circular logic. It works because it assumes itself.

Now ask the deeper question:

"Where is this inertial frame located?"

The answers will become uncomfortable. They'll point to deep space, or to idealized reference points, or even suggest Earth itself — with a smirk and a shrug.

What no one will say directly is the truth we're exposing now:

Every practical application of physics treats the Earth as an inertial frame — even while claiming it's not. This is the

Inertial Frame Illusion — the quietest hypocrisy in the foundations of modern science.

What Is an Inertial Frame, Really?

By textbook standards:

- An inertial frame is a reference frame where an object not acted upon by a net external force moves in a straight line at constant velocity.
- In such a frame, Newton's laws work without correction.

But let's go deeper. How do we test if a frame is truly inertial?

There is no cosmic "still point" where you can anchor your instrument. You cannot step outside the universe and check. So, in practice, physicists do the next best thing:

They use the Earth as the default inertial frame — and then apply "corrections" (Coriolis, centrifugal, relativistic) when those approximations fail.

But here's the paradox:

- If Earth is *not* an inertial frame (because it spins and orbits), then Newton's laws shouldn't strictly apply.
- But we use them anyway. Daily. In labs. In rockets. In simulations. In aircraft.

So what's happening?

The answer is simple but rarely spoken:

Every known reference frame is only valid when it quietly assumes Earth is still.

The Great Disguise

When Newton formulated his laws, he never stated Earth is perfectly inertial. But he built a world where physics *behaves* as if it is.

Why?

Because Earth's stillness is necessary for predictability.

If Earth's spin and orbit were constantly felt, measured, and adjusted for — the mathematics of motion would become chaotic. The laws would become unstable. Engineering would become impractical.

So, the scientific community made a convenient compromise:

- Assume the Earth is still for calculations.
- Assume the Earth is moving for cosmology.
- But never hold both assumptions in the same room.

This is not science. This is strategic storytelling.

Frame of Convenience, Not Truth

From classical to quantum to relativistic physics, reference frames are the invisible scaffolding of all equations.

But ask yourself:

- Why does a pendulum behave as expected in your lab?
- Why do Newton's equations work for your projectile?
- Why does a car accelerate in a straight line, and a drone hover steadily in one place?

Because you assume — either explicitly or implicitly — that the Earth under your feet is stationary.

You treat Earth as still to make the math work. And yet, you say Earth moves because the models require it cosmologically.

This split loyalty is the illusion.

UED's Fracture Line: Stillness is the Default

Under Universal Energy Dynamics (UED), this contradiction dissolves.

UED says:

"Stillness is not an approximation — it's the foundational state of the structured vacuum field. All observed motion is a ripple imbalance resolving through local participation. The observer's frame — Earth — is not in motion, but hosting ripple interactions."

That means:

- The reason Newton's laws work on Earth is not because Earth is "almost" inertial.
- It's because Earth actually *is* the local still field in ripple equilibrium.
- And inertial behavior is simply the appearance of ripple alignment in that field.

There's no need to correct for spin, orbit, or movement — because those are interpretative overlays, not physical truths.

A Thought Experiment: The Spinning Lab

Imagine you're conducting a physics experiment on a spinning turntable. You drop a ball — it deflects sideways.

Now, someone outside the turntable says, "That's not a real force — it's a pseudo-force caused by your rotation." You correct your equations accordingly.

But now consider this:

What if you're on Earth — and Earth is also spinning? Shouldn't we correct *everything*?

We don't. We only apply corrections in specific geophysical cases — missile launches, weather systems, satellites. Even then, we patch the equations with "non-inertial corrections" — but we don't change the fundamental framework.

That's the illusion.

We're pretending to honor Earth's motion...While actually relying on its stillness.

The Dirty Secret of All Physics

Here's the uncomfortable truth:

The entire edifice of experimental physics is built on the assumed stillness of the Earth.

No one says it out loud — but every lab, every machine, every controlled setup quietly leans on the idea that:

- There is a reliable "down."
- There is a consistent "rest."
- There is a definable "direction" that isn't rotating beneath your feet.

Remove that foundation, and even your simple pendulum becomes a chaos simulator.

Final Thought for This Chapter

The inertial frame illusion is not a malicious lie. It is a survival strategy of a system that *needed stillness* but didn't have the courage to admit it.

We have no problem saying "the Sun appears to rise" — because we believe we've upgraded the truth.

But we are still anchoring our physics in a still Earth — all while denying it.

UED simply reclaims the obvious:

Stillness is not a workaround. It's the real frame. Earth doesn't move — because it doesn't need to.

What moves is the field, the ripples, and your beliefs.

And now... your frame of thinking is shifting too.

4: The Constant Arrow of Time

UED View: Time as Threshold of Ripple Perception

We are told time flows like a river — forever forward, never backward. This metaphor is repeated in every culture, every classroom, and every calendar. But no one asks: where is this river flowing, and what makes it move?

We wear watches. We chase deadlines. We age. We count seconds, minutes, hours — convinced they're real.

Yet if you stop for a moment, close your eyes, and strip away your schedules — what do you *actually* experience?

Do you feel time... or do you feel change? Do you observe time... or do you observe consequences?

The answer is subtle but profound:

Time is not a thing — it is your interpretation of ripple transitions through structured stillness.

This chapter will dismantle the illusion of the "constant arrow of time" and replace it with the UED principle:

Time is the threshold of perception that emerges when ripples interact, resolve, and align within your consciousness.

What Is the Arrow of Time?

Conventional physics teaches us that time has a direction — from past to future — and that it flows *uniformly*.

This belief is anchored in two main pillars:

1. **Thermodynamics:** The second law says entropy always increases. Therefore, time must move forward.
2. **Relativity:** Time can stretch or compress depending on speed and gravity, but its arrow still points forward.

This idea is so entrenched that even our language and logic depend on it:

- We speak of causes preceding effects.
- We plan for the future and regret the past.
- We believe "before" and "after" are embedded into the very fabric of the universe.

But what if they're not?

What if time doesn't flow? What if it emerges only when ripples align in a certain way — and disappears when they don't?

UED Disruption: Time Isn't Flowing — You Are Observing

Under Universal Energy Dynamics (UED), the universe is not a clock. It is a still structured field in which energy, intent, and consequence ripple across equilibrium points.

When we perceive time, we are not witnessing some invisible line stretching into the future. We are:

- Perceiving ripple transitions from one resolved consequence to another.
- Noticing thresholds where new information becomes visible.

- Calling that unfolding: time.

But the reality is deeper:

Time is not a river. It's a ripple ripple ripple. It is the illusion of continuity stitched together by memory and triggered by change.

No ripple → no change → no new perception → no time.

This is why in sleep, coma, or deep meditation, time disappears. And in trauma or high focus, it dilates. Not because your clock changed — but because your threshold of ripple perception shifted.

ATT – The Alignment Threshold of Time

UED introduces a new concept: ATT (Alignment Threshold of Time).

This is the point at which ripple interaction reaches perceptual alignment, triggering the emergence of a temporal sense.

You don't see "time" pass. You see events emerge once they cross your ATT — the threshold of your field's ability to resolve interaction.

It's like a ripple sonar:

- When ripples align with your perceptive range → time feels "normal."
- When ripple frequency increases → time feels "fast."
- When ripples misalign → time vanishes or stretches.

This is not psychological. It's field-energetic.

Practical Examples of Ripple-Time

- **Dreams:** Entire narratives unfold in seconds. Why? Because ATT is bypassed — the ripple chain is compressed.
- **Meditation or Samadhi:** Time ceases to exist. Ripple noise is silenced. No new thresholds crossed → no perception → no time.
- **High-Speed Accidents:** Time slows down. ATT increases due to shock; every micro-ripple becomes accessible → time expands.
- **Deep Focus (Flow):** Ripple noise reduces, alignment becomes tight → actions seem timeless.

All these are modulations of ATT, not distortions of some ticking cosmic clock.

The Fallacy of the Constant Arrow

The idea that time flows constantly, for everyone, is a product of industrial thinking:

- We needed synchronized labor, machines, schedules.
- We engineered clocks, time zones, world standards.
- And we began believing in mechanical time over experiential consequence.

But nature doesn't care for clocks.
In fact:

- A tree doesn't grow in seconds — it grows in seasons, ripples, responses.

- A fetus doesn't calculate weeks — it ripples into life through field interaction.
- A black hole doesn't respect seconds — it collapses ripple tension, swallowing time itself.

So ask:

If time is so constant, why can it be stretched, bent, lost, or paused?

Because it's not fundamental. It's perceived rhythm, not absolute reality.

Time Is the Shadow of Consequence

Let's go philosophical for a moment:

When nothing happens, you feel timeless. When everything happens too fast, you feel timeless. Time only emerges when consequence unfolds in observable sequence.

That's all time is: A shadow cast by the movement of ripples across awareness.

It's not linear. It's not constant. It's quantized emergence of perceptual change.

This is why causality, not time, is the real player. But even causality is ripple-based, not timeline-based.

Final Thought for This Chapter

We've built entire cosmologies on the back of time — its passage, its beginning, its end.

But UED cuts through the illusion:

Time is a perceptual threshold. It begins when ripples align. It ends when consequence is resolved. It never flows. You do.

You, the observer, move through ripple zones. You cross thresholds of awareness. You perceive motion, emotion, evolution — and you call that *time*.

But the universe? It remains still. Timeless. Waiting only for your next ripple to arrive.

And when it does — you'll say, "Time has passed." But in truth... only you moved.

5: Force: A Symptom, Not a Cause

Deriving force as imbalance-seeking resolution

The concept of force is central to all of physics.

From Newton's $F = ma$, to the pushing of a door, the thrust of a rocket, or the gravitational pull of planets—force is seen as the primary cause of motion and change.

But what if this entire framework is backward?

What if force is not the cause of motion at all—but merely a symptom of something deeper, something more foundational?

What if force, as we know it, is just a visible ripple struggling to restore equilibrium?

That's the bold claim of Universal Energy Dynamics (UED):

Force is not a cause. It is a consequence. A ripple expression of field imbalance. Whenever force appears, something is already out of harmony—and the force is the field's attempt to resolve it.

The Classical Illusion: Force as the Mover

In Newtonian physics, force is fundamental:

- Apply a force, and you get acceleration.
- No force, no acceleration.

- Force is a direct agent of change.

This thinking has served us well—from levers to rockets to bridges.

But it also created a mechanical worldview, in which the universe is seen as a giant machine, ticking because something pushes, pulls, or presses upon something else.

In this worldview:

- Force initiates.
- Objects obey.
- Reality moves.

But there's a silent assumption in this entire structure:

That the universe begins in rest, and force must act upon it to disturb or shift that rest.

UED exposes the paradox:

If rest is the natural state, and force is needed to change it, where does the first force come from? Who applied the very first push, the first pull?

This leads to a foundational inconsistency in classical physics—where motion is caused by something that itself has no origin.

And so begins the unraveling.

UED Reframe: Fields First, Not Forces

Universal Energy Dynamics starts from a very different first principle:

The universe is fundamentally still. Every object, phenomenon, or particle is a ripple disturbance in a structured vacuum energy field (SVE). These ripples are not objects in motion, but imbalances in field symmetry.

From this lens:

- Force is not what causes motion.
- Force is what emerges when equilibrium is disturbed.

It is not a creator of change. It is a signal that change is trying to resolve.

Just like:

- A fever is not the disease—it's the symptom.
- A crack is not the stress—it's the result of stress seeking release.
- A wave crashing on the shore is not force—it's energy trying to resolve a boundary imbalance.

The True Sequence (UED View)

Let's compare Newtonian and UED sequences:

Traditional Physics	UED Framework
Force → Motion	Field Imbalance → Ripple Participation → Apparent Motion → Force as Surface Expression
Force causes acceleration	Acceleration appears due to unresolved ripple tension
Force is active	Force is reactive

In UED, the sequence looks like this:

1. A field equilibrium is disturbed (by intent, interaction, or boundary constraint).

2. A ripple is triggered to rebalance the field.
3. This ripple interacts with structures (like particles or bodies).
4. The observable effect on these structures is what we call force.

So, force is not the start of the process—it's the visible midpoint in a deeper resolution loop.

Everyday Analogy: The Ripple in Water

Imagine a calm pond. You place your finger gently in the water and move it. Ripples emerge—not because water "decides" to ripple, but because the equilibrium was disturbed.

The water is not "acting" with force. It's responding to imbalance. Now scale this up:

- What we call "pushing a box" is actually creating a ripple mismatch between your energy field and the box's field.
- The box "moves" because ripple resolution requires displacement toward equilibrium.
- The force you "feel" is feedback from the unresolved tension.

You are not *causing* the motion. You are participating in a ripple loop where force is the feedback signal—not the initiating agent.

Gravitational Force: A Case Study

Gravity is always cited as the most intuitive force—"what goes up must come down."

But what is gravity in UED terms?

- It is not a pulling force from the Earth.
- It is the ripple contraction of field tension caused by condensed energy (mass) in a structured vacuum.

Objects fall not because they're pulled, but because they seek equilibrium in the ripple compression field caused by Earth's presence.

Thus:

- The apple doesn't "fall" due to force.
- It aligns with downward ripple gradients until stillness is restored.

This is ripple resolution—not attraction.

The Ripple Equation of Force (Conceptual)

If we were to write a symbolic logic of force in UED, it might look like this:

$F \propto \Delta R / \Delta t$

Where:

- F = Emergent force (felt effect)
- ΔR = Change in ripple alignment (tension shift)
- Δt = Perceptual threshold time (not absolute time, but ATT-based)

This frames force not as a push, but as rate of change in ripple misalignment, relative to the observer's field sensitivity.

When ΔR is high and Δt is low (rapid, strong ripple conflict), the force feels "violent" (impact, collision). When ΔR is gentle, the force feels soft (gravity, buoyancy). When ΔR resolves slowly, force becomes tension, pressure, or even intention.

Philosophical Implication: Stillness Is Not Weakness

If force is a symptom, then stillness is not the absence of energy. It is the completion of ripple resolution. It is fulfilled alignment. A peaceful state is not a passive state. It is the consequence of resolved field dynamics.

This reframes everything from physics to emotions:

- Anger is unresolved ripple pressure.
- Movement is ripple imbalance.
- Force is feedback.
- Peace is field coherence.

Final Thought for This Chapter

Newton taught us that force moves objects. But UED shows us that objects don't move at all. They ripple. And force is simply what you feel when the ripple wants to resolve — but hasn't yet.

Force is not a cause. It is the field's way of showing you that equilibrium has been disturbed. The next time you push, pull, or fall—don't say, "I applied a force." Say,

"I entered a ripple conflict." And the universe responded.

6: Mass and Matter

From Intrinsic to Condensed Ripples

Why mass is not a built-in property but an energetic consequence

We are taught that mass is one of the most fundamental properties of matter.

An object has mass. Mass resists motion. Mass attracts other mass. Mass curves space. Mass converts into energy (and vice versa).

In short, mass is presented as an inherent, inescapable trait — something a particle simply "has" by nature.

But let's stop and ask:

What is mass... really? Is it something that sits inside an object like a hidden weight? Or is it simply how that object behaves under certain ripple interactions?

Under Universal Energy Dynamics (UED), mass is not intrinsic. It is not "carried" by a particle. Instead, mass is:

The energetic consequence of ripple condensation in a structured still field. It is the field's response to localized alignment resistance — not a property of the object itself.

This chapter rewrites one of the most sacred assumptions of physics — by showing that mass is not a thing, but a resolved condition.

The Classical Illusion: Mass as Built-in Identity

In Newtonian mechanics:

- Mass is a scalar quantity.
- It defines how much force is needed to accelerate an object.
- It is considered constant, absolute, and inherent.

Later, Einstein's relativity made mass even more mystical:

- It "increases" with speed.
- It can warp spacetime.
- It is equivalent to energy via $E = mc^2$.

Yet in all these views, mass is treated like a core identity — something that "belongs" to the object, even if it can shift in form or perception.

But this idea has never been fully explained.

- *What gives mass its value?*
- *Why does an electron have its specific mass?*
- *Why does light have no rest mass?*
- *Why can mass convert to energy—but energy doesn't always condense into mass?*

These are not answered by classical models. They are merely wrapped in mathematical definitions or speculative quantum fields (e.g., Higgs field), which shift the question, not solve it.

UED brings clarity.

Ripple Condensation: The UED View of Mass

According to Universal Energy Dynamics, the universe is fundamentally a structured vacuum energy field (SVE) — a still, intelligent field where ripples arise due to intent, imbalance, or interaction.

Mass appears when ripples condense and stabilize, resisting further alignment with surrounding equilibrium.

Let's simplify it:

- When ripple waves pass through each other without resisting, no mass appears. → This is like light.
- When ripple waves condense, knot, or oscillate in stable loops, resisting further resolution, a localized tension builds in the field.
- This resistance to further ripple realignment is what we call mass.

Thus: Mass is the measurable outcome of structured ripple resistance.

It is not a "thing" but a state of tension held in place — like a whirlpool in water, stable but never static.

Matter as Standing Ripple Structures

Matter, then, is simply: An organized structure of condensed ripples stabilized in local equilibrium zones.

Atoms, electrons, protons — these are not solid "balls" of stuff.

They are persistent ripple knots that hold their structure against the surrounding stillness.

In UED, what we call "particles" are:

- Ripples with boundary behavior.

- Field oscillations stabilized by geometric or energetic feedback.
- Temporarily localized memory of imbalance, embedded in space.

They behave like matter because:

- They condense energy.
- They resist ripple penetration.
- They store pattern.

This means matter is not a different "substance" from field — it's just a more stubborn ripple alignment.

Why Mass Varies

In UED, the mass of a particle (say, an electron) is not handed to it by some universal deity or hidden Higgs field.

Instead, its mass is determined by:

- The density of its ripple condensation.
- The symmetry resistance of its surrounding field.
- The stability of its oscillation loop.

So an electron has a certain mass because:

- Its ripple formation reaches a threshold that resists being undone.
- It reaches minimum ripple stability for continuous appearance.
- It locks into field memory — forming a localized persistence.

Different particles = different ripple geometries = different resistance = different mass.

It's not mystical. It's ripple geometry and field tension.

Why Light Has No Mass (but Still Exists)

Light, or photons, are ripple participants too — but they behave differently.

They do not condense into a localized field knot. They don't resist ripple alignment. They pass through equilibrium without embedding.

Thus:

- They carry energy (ripple intent).
- They express frequency.
- But they do not manifest resistance, so they appear massless.

They are ripple events, not ripple containers.

Massless, yet real. A ripple — not an object.

Why Gravity Affects Mass

In Newtonian and Einsteinian views, gravity acts on mass. But neither fully explains why mass "feels" force.

In UED:

- Gravity is a ripple pressure gradient — a curvature of ripple density toward a field condensation (Earth, Sun, etc.).
- An object with more condensed ripples (i.e., more mass) interacts more stubbornly with this gradient.
- Therefore, it appears "heavier."

Heaviness is not because of attraction. It is the friction of ripple resistance in a field gradient.

And thus, gravity's interaction with mass becomes obvious:

It's not pull. It's pressure on persistent ripple tension.

$E = mc^2$ Reframed

Einstein showed that mass can become energy — and vice versa.

UED explains why:

- Mass is energy — but in structured, condensed ripple form.
- When released (via fusion, fission, annihilation), these structures break their ripple knots, liberating their stored oscillation back into unstructured ripple energy.

Thus, energy is not something else. It's just mass unknotted.

And the speed of light squared (c^2) is the ripple-permissible intensity of energetic emergence into perceived field consequence.

Philosophical Implication: You Are Ripple Tension Too

If all mass is ripple memory, and all matter is condensed consequence — then what are you?

You're not a solid object.

You are:

- A walking ripple knot.
- A temporarily stabilized pattern of structured

resistance.

- A conscious condensation of stillness — experiencing itself through time, force, and consequence.

You're not made *of* matter. You are made *by* ripple tension.

Your body. Your mind. Even your thoughts — All are varying densities of ripple resistance.

The universe isn't a material machine. It's a still song, condensed into rhythm.

Final Thought for This Chapter

Mass isn't real in the way we were taught.

It's not a nugget of "stuff" lodged in particles. It's not granted by mysterious fields. It's not a divine property written into physics.

Mass is the fatigue of ripples refusing to let go. It's the echo of imbalance that refuses to dissolve. It's the tension of stillness held too tightly.

Matter is not the substance of the universe. It's the symptom of a field choosing to remember.

And you — as a ripple being — are one such memory.

Interlude: The Stillness Beneath the Storm

Six chapters. Six sacred assumptions dismantled. Not with rebellion, but with resolution. Not by denying science, but by reminding it where it came from.

You've just walked through the quiet unraveling of the loudest pillars in classical physics:

- A motionless Earth called *stationary* not by ignorance—but by precision.
- Inertia, not as an object's stubbornness, but as ripple equilibrium waiting to resolve.
- Frames of reference, not as neutral perspectives, but disguised confessions of stillness.
- Time, no longer a flowing arrow, but a perceptual ripple threshold.
- Force, not a cause, but a symptom.
- And mass—not matter itself, but matter remembered by the field.

What you've seen is not destruction. It's calibration.

You've been invited to hold two perspectives:

- The one you inherited: mechanical, fragmented, linear, imposed.
- And the one you're now uncovering: ripple-based, coherent, causal, still.

In **Part I**, we broke the illusion. In **Part II**, we begin to build the new lens.

Because now that we've cleared the fog of assumed motion, constant time, and built-in properties— We're ready to ask:

If the Earth doesn't move, then how do we explain day and night? If time is not real, how do rhythms emerge? If mass is ripple resistance, then what is gravity?

Welcome to the ripple. Where stillness is truth, and everything else is consequence.

Let the reconstruction begin.

Part II – The Moving Earth Myth

7: Spin Deception

Day and Night Without Rotation

Observational perception vs. assumed motion

Every morning, the Sun rises from the east. It arcs across the sky, lighting up the world. By evening, it sets in the west. Darkness falls. The Moon appears. Then comes night.

We've witnessed this cycle since the dawn of human awareness. It's simple, rhythmic, elegant.

And yet—somewhere along the historical timeline—we were told:

"No, the Sun isn't moving. *You* are."

This single statement flipped the script of cosmology.

- What you *see* is wrong.
- What you *feel* is misleading.
- What you *observe* is illusion.

Instead, trust the mathematics. Trust the model. Trust the spin.

And so, we began teaching that Earth spins eastward at 1,674 km/h—even though nothing in your daily experience remotely suggests it.

This is the Spin Deception:

The replacement of direct observation with an assumed mechanism that has never been physically felt, tested, or independently confirmed without embedding assumptions of motion.

The Root of the Deception: Trusting Models over Senses

Why do we believe Earth is spinning?

Because we see:

- Day become night.
- Stars appearing to revolve.
- Sun rising and setting.

But instead of accepting that these *appear* as direct celestial motions, we've been taught to reverse the interpretation.

We're told:

- "It's not the Sun moving. It's the Earth spinning."
- "It's not the stars revolving. It's the sky appearing to move due to Earth's rotation."

This reversal is not based on direct evidence. It's a model-based explanation — inherited from the Copernican revolution and reinforced by Newtonian mechanics and relativity.

The observational evidence is neutral:

- The Sun moves in the sky.
- The Moon follows a path.
- Stars rotate around Polaris.

But instead of asking what we're actually seeing, the model presumes a spinning Earth, and explains all visible motion from that foundation.

It's not physics. It's interpretation.

The Invisibility of Spin

Let's ask the obvious question:

If Earth is spinning at over 1,600 km/h, why don't we feel it?

The conventional answer is:

- "Because everything moves with it, uniformly. It's like being on a smooth train."

But even a train, no matter how smooth, reveals its motion through:

- Acceleration and deceleration
- Curvature
- Vibrations
- Wind resistance when standing outside

Yet on Earth:

- You can balance a pencil on its tip.
- You can jump straight up and land on the same spot.
- Birds can hover, insects can float, drones can hang still.

If Earth were truly spinning at high velocity:

- Why is the air not rushing violently westward?

- Why can weather systems move in all directions?
- Why do airplanes fly east and west at similar durations?

These aren't trivial questions. They expose a deeper issue:

The spin we're taught to believe in has no directly observable evidence — yet it overrides our lived perception.

This is philosophical gaslighting in scientific form.

The UED View: Perception Is Consequence, Not Cause

In Universal Energy Dynamics, nothing moves in the classical sense. There is no real motion — only ripple consequences in a still, structured field.

From this perspective:

- The Sun's rising and setting is not caused by Earth spinning.
- It is a rhythmic alignment of ripple perception based on your position, your consciousness, and local field geometry.

The day-night cycle is a resolved pattern—a perceptual rhythm that emerges as the observer moves through zones of field alignment.

Think of it like a ripple clock:

- Light isn't *traveling* to you from a rotating position.
- It is activating visibility in your location as your local ripple field reaches threshold (ATT — Alignment Threshold of Time).
- When alignment fades, perception collapses, and we

call that "night."

So, the Sun's movement is not a celestial motion—but a consequence of ripple field resonance activating in sequential coherence.

The result looks like movement. But the source is stillness.

Day and Night Without Rotation

Let's now break the key idea down:

What You Observe:

- The Sun "rises" and "sets."
- Shadows move.
- Light transitions from darkness to brightness and back.

What You Were Told:

- This is due to the Earth spinning eastward.
- You're on a rotating sphere, and that rotation causes the apparent movement of the Sun.

UED's Explanation:

- Earth is still.
- The structured vacuum field around Earth has zones of ripple tension that permit or block solar resonance visibility.
- As your location moves in alignment with the ripple structure, you see the Sun.
- As it moves out of alignment, the Sun disappears — not because it moved, but because your field stopped

resolving it.

It's not *sunlight moving.* It's visibility activating and deactivating in rhythmic resonance.

Sun as a Signal, Not a Body in Transit

UED treats the Sun not as a giant burning ball moving through space, but as a cosmic field activator — a perceptual source, always in energetic presence, but seen only when ripple alignment resolves to threshold.

Think of a Wi-Fi signal:

- It is always present in the air.
- But your phone shows "connected" only when resonance aligns with the field strength.

Likewise, the Sun is always "there," but it becomes visible only through ripple participation that activates light perception in your field.

Thus:

- Sunrise = Field alignment.
- Noon = Peak ripple visibility.
- Sunset = Alignment loss.
- Night = Dormant signal perception.

And Earth? Earth didn't spin. You moved through resonance.

Why the Sky Appears to Rotate

The same explanation extends to stars and celestial objects.

They are not moving around the Earth. They are stationary field signatures, activated through rotational alignment of perception, much like a cosmic kaleidoscope.

As you turn your awareness (geographically and energetically), different parts of the still field ripple into your perceptual zone.

You say, "the sky is moving." UED says, "your ripple window is rotating."

Again, the cause isn't spin. The cause is your participation in layered field visibility.

Final Thought for This Chapter

The Earth doesn't spin. It doesn't need to.

Day and night arise not because the planet is a ball on a cosmic treadmill, but because ripple-based perception moves through structured resonance zones in the still universal field.

Motion is an illusion of the unresolved. The Sun is not moving around you — nor are you spinning toward it.

You are simply living inside a rhythm of awareness, watching light unfold like clockwork from a place that never needed to move.

This is not deception. This is deeper precision.

8: Orbit Fallacy

Does the Earth Really Revolve?

Field alignment vs. heliocentric storytelling

We are told a story from childhood— A majestic one, taught in every classroom, illustrated in every science book:

"The Earth revolves around the Sun in a perfect elliptical orbit, traveling at nearly 108,000 kilometers per hour."

This idea is so deeply embedded that even questioning it feels like heresy. We repeat it with pride, model it with thermocol balls, and even measure our calendars around it.

But how do we know it's true?

- Have we seen the Earth move through space?
- Have we directly measured the 108,000 km/h revolution?
- Do we observe the orbit, or do we just interpret seasonal changes as orbital consequences?

This is the Orbit Fallacy—the belief that the Earth's revolution is a physical truth, rather than a narrative built on assumptions, stitched together by mathematical convenience and blind trust in heliocentric storytelling.

Under Universal Energy Dynamics (UED), we dismantle this illusion—not by denying the rhythm of seasons or celestial consistency, but by showing that orbit is not motion, but ripple field realignment.

The Heliocentric Narrative: A Grand Assumption

Let's first revisit the standard view:

- The Earth revolves around the Sun once every 365.25 days.
- This revolution is responsible for the changing seasons.
- The orbit is elliptical, with the Sun at one focus.
- This model replaced the geocentric (Earth-centered) model of ancient times.

While this heliocentric model improved mathematical predictions of planetary positions, it came with a hidden assumption:

That revolution must mean physical motion.

In other words, because we observe periodic celestial changes, we assume the Earth must be physically moving through space to account for them.

But is that necessarily true?

Could those changes arise from something deeper, more subtle—like field alignment cycles rather than planetary locomotion?

Let's explore.

What Do We Actually Observe?

Let's strip it down to raw observation:

- We see the Sun's position in the sky change throughout the year.
- We notice different constellations becoming visible in different seasons.
- We feel temperature shifts—summer, winter, spring, and fall.

Do these *prove* the Earth is orbiting?

No.

They prove that something in our perception of the celestial field changes cyclically. But that doesn't confirm a 108,000 km/h orbital motion. That is an interpretive framework layered over field rhythms.

Just like sunrise and sunset can be explained without spin, seasonal and stellar changes can be explained without orbit—if we view them as field resonance shifts, not mechanical planetary motion.

UED View: Field Alignment, Not Orbital Travel

According to Universal Energy Dynamics, all motion is illusory. What we perceive as motion is actually:

A ripple response through structured vacuum field (SVE), where perception emerges at zones of alignment.

From this lens:

- The Earth doesn't need to physically revolve.
- What changes is our alignment with the surrounding ripple field, which activates different solar and stellar visibility patterns.

Think of it this way:

- You are standing still inside a massive, structured field.
- As your field resonance shifts over time, different celestial elements become perceptible.
- This creates the illusion of orbit, but nothing has moved in space—only field harmonics have changed.

In other words:

"Orbit" is just rhythmic realignment within a stable field matrix.

Earth as a Fixed Point of Perception

In the heliocentric model, the Earth is a mobile observer—racing around the Sun, gathering different views as it goes.

In UED, the Earth is a fixed field anchor—an observer embedded in stillness, watching ripples of visibility activate and deactivate around it.

- The Sun doesn't move.
- The Earth doesn't move.
- Perception moves—and that movement is a function of ripple coherence within your field.

This explains:

- Constellations rotating throughout the year — not because Earth is orbiting, but because the field around Earth is rotating ripple alignments.
- Seasonal light variations — not due to axial tilt and orbit, but due to field density shifts in the solar ripple

zone.

Thus:

- Summer = Stronger alignment with high-energy solar ripple zones.
- Winter = Weaker alignment due to detuned ripple coherence.

No orbit needed. Just field realignment.

A Thought Experiment: The Sun in a Ripple Sphere

Let's visualize it:

Imagine the Sun surrounded by a giant, structured ripple sphere.

This ripple sphere has zones—bands of energetic tension, alignment nodes, interference patterns.

Now imagine Earth is not moving at all—but as time progresses:

- The ripple sphere subtly rotates or oscillates,
- Causing Earth's local field to align differently with the solar resonance.

This creates:

- The appearance of orbit,
- The appearance of changing constellations,
- The rhythm of seasons,
- Without any actual planetary motion.

In this model:

Orbit is a ripple illusion caused by spherical field rotation, not physical displacement.

Challenging Kepler, Completing Copernicus

Kepler gave us ellipses. Copernicus gave us the shift from Earth-centered to Sun-centered. Einstein curved it all.

But none of them asked:

- *What is the Sun embedded in?*
- *What defines the Sun's own field?*
- *If everything is relative, then why is Earth's orbital velocity considered absolute?*

UED re-centers the conversation:

- It's not about who moves around whom.
- It's about how fields respond to energy condensation.

Just as electrons don't *orbit* nuclei in UED (they resonate in fields), Earth doesn't orbit the Sun—it resonates with solar ripple fields.

It's not celestial ballet. It's ripple choreography.

Practical Implications

If Earth doesn't revolve:

- There is no centrifugal force from orbit.
- The solar system is not a swirling machine—it's a field mosaic.
- Our calendars become alignment trackers, not planetary stopwatches.
- GPS, astronomy, and satellite systems may still *work*, but their models are approximations, not truths.

This doesn't invalidate technology. It reveals the illusion beneath the code.

Final Thought for This Chapter

We've built a story around motion—because motion was all we knew.

But motion is just the skin. The muscle is ripple response. The skeleton is stillness.

The Earth doesn't revolve. It listens.

And what we call an orbit is nothing but the echo of ripple harmonics returning to coherence.

So the next time someone says,

"The Earth orbits the Sun," You can quietly smile and say, "Only in the story. In the field, it realigns."

9: Gravity: From Attraction to Ripple Pressure

The UED alternative to the Newton–Einstein gravity myth

No force has held more power over the scientific imagination than gravity.

- Newton called it an invisible force of attraction between masses.
- Einstein reimagined it as the curvature of spacetime.
- Quantum theorists try to fit it into the particle zoo (with the hypothetical *graviton*).
- Engineers use it every day to predict planetary motion, launch rockets, and build bridges.

And yet, gravity remains the least understood of all the fundamental forces.

It doesn't fit comfortably into quantum mechanics. It resists unification. It has never been directly observed—only inferred. And most importantly, its cause remains a mystery.

Both Newton and Einstein described what gravity *does*, but neither could explain *why* it exists in the first place.

In Universal Energy Dynamics (UED), we challenge this foundational myth. We do not treat gravity as attraction or as curved space, but as something more fundamental:

Gravity is ripple pressure—an expression of tension resolution within the Structured Vacuum Energy (SVE) field caused by the presence of condensed energy (mass).

It is not a pull. It is not a warp. It is a field's intelligent response to imbalance.

Let's reframe one of the most sacred concepts in physics—through stillness.

Newton's Gravity: Force Without Mechanism

Newton's law of gravitation tells us:

Every mass attracts every other mass with a force directly proportional to their masses and inversely proportional to the square of the distance between them.

$F = G * (m_1 m_2 / r^2)$

This equation is elegant. It works. But what causes this attraction? Newton himself admitted:

"I frame no hypotheses."

He didn't explain why gravity works. He simply noticed that it appears to behave consistently—and defined it as a mutual pull between objects.

But this view carries several assumptions:

- That mass is intrinsic.
- That forces act across empty space.
- That attraction happens instantly (or nearly so).

UED questions all of these—and replaces them with field-based dynamics.

Einstein's Gravity: Geometry, Not Force

Einstein improved upon Newton by proposing that:

Mass doesn't pull—it curves spacetime. Objects follow "geodesics" (natural paths) in this curved geometry.

This eliminated the need for force-at-a-distance. Instead of being "pulled," objects fall because they follow the shape of the spacetime fabric.

But again—what is this fabric? What curves it? And what *is* spacetime, fundamentally?

It's never clearly defined. Instead, we rely on analogies:

- Stretch a rubber sheet.
- Drop a bowling ball.
- Watch the marbles roll inward.

But these analogies raise deeper questions:

- What supports the sheet?
- What causes the sheet to curve?
- What is being stretched—matter? Energy? Space itself?

Einstein offered description, not origin.

UED Breakthrough: Gravity Is Not Attraction—It Is Ripple Pressure

In Universal Energy Dynamics, we abandon both Newton's pull and Einstein's warp. Instead, we define gravity as:

A pressure differential in the structured vacuum field caused by ripple imbalance around a condensed energy zone (i.e., mass).

Let's break that down:

- Mass, in UED, is not a substance. It is a localized ripple condensation—a persistent knot of energy resisting field equilibrium.
- This condensation causes tension in the surrounding field—like a depression, but not in space, rather in ripple alignment.
- The SVE field, being inherently intelligent and still, responds by trying to resolve this tension.

This response appears as a force—but it is not a pull. It is a field pressure gradient, guiding nearby ripples (and objects) toward lower tension zones, i.e., toward the mass.

Thus: Gravity is not attraction between masses. It is a ripple field's push toward restored balance.

Visualizing Ripple Pressure

Imagine a calm, tensioned sheet of water—perfectly still. Now, drop a pebble (mass). Ripples emerge.

These ripples are not "pulling" anything—but they create pressure gradients that draw lighter objects inward, not due to force, but due to field pressure differences.

In space, Earth is the pebble. Surrounding field tries to restore stillness. Anything in the zone of ripple tension feels pressure inward—and we interpret this as "gravitational pull." But it's not pull—it's suctionless participation in field correction.

Why Objects Fall

In Newton's view: The Earth pulls objects downward with gravitational force.

In Einstein's view: Objects follow curved spacetime downward.

In UED: An object placed above Earth's surface enters a ripple field already under tension. The object's own ripple structure participates in resolving this imbalance. It descends as a ripple-consequence seeking equilibrium—not because of "mass," but because of misaligned field pressure.

There is no force, only pressure differential. And this explains why:

- All objects fall at the same rate (ripple equilibrium seeks resolution independent of mass shape).
- Free fall feels weightless (because the object is fully immersed in resolution, with no counter-ripples).
- Gravity is weakest far from condensed energy zones (where field tension has dissipated).

What About Orbits?

In classical physics, orbits are balances between gravitational pull and inertia.

In UED:

- A satellite is a ripple-condensed object suspended at a harmonic node in the Earth's field.
- It does not "fall around the Earth."
- It resonates with the ripple pressure gradient, staying suspended as long as its ripple momentum balances the local field tension.

Change the field (drag, decay, collision), and it descends—not by pull, but by loss of field fit.

Black Holes and Gravitational Collapse

Black holes are the ultimate gravitational phenomena in relativity.

But UED sees them differently:

A black hole is a point of infinite ripple tension—where all field resolution fails, and energy collapses into a condensed ripple singularity.

Nothing escapes, not because gravity pulls it in, but because the ripple tension is too high for re-alignment. In this zone:

- Light fails to align (hence "darkness").
- Mass cannot stabilize.
- Time perception disintegrates (ATT collapse).

Again, this is not warping of space, but breakdown of ripple coherence.

Why This Matters

- It replaces "gravity" with field behavior.
- It unifies motion, mass, and force under one ripple model.
- It removes the need for:
  - A pulling force (Newton)
  - A mystical spacetime fabric (Einstein)
  - A hypothetical graviton (quantum theory)
- It allows energy, awareness, and matter to be seen as ripple stages in a coherent field.

Gravity is no longer mysterious. It is the universe's reflex to preserve stillness.

Final Thought for This Chapter

We've spent centuries believing in attraction.

We called it love when it was survival. We called it gravity when it was balance-seeking. We called it force when it was ripple pressure.

Now we know better.

The Earth doesn't pull you. It invites your ripples into resolution. And you fall—not down, but back into balance.

That's not gravity. That's truth remembering stillness.

10: Why You Don't Feel the Spin

Real stillness and the illusion of reference shift

You've been told your entire life that you're spinning — fast.

At the equator, Earth rotates at 1,674 km/h. It orbits the Sun at 108,000 km/h. The Sun orbits the galactic center at 828,000 km/h. And the galaxy itself is racing through the universe at over 2 million km/h.

According to mainstream cosmology, you are a passenger on a cosmic roller coaster that never slows down.

Yet here's the contradiction:

You feel perfectly still.

- You can balance a coin on a table.
- A candle flame burns straight upward.
- You sleep peacefully.
- You look at the Moon — and it appears as if you are not moving at all.

How can this be?

Is your body just numb to motion? Have your senses evolved to ignore extreme velocities? Or is the truth much simpler — and far more revolutionary?

This chapter explores the illusion of reference shift, and how true stillness — not hidden movement — is what anchors all experience.

The "Spin" Is Not Observable — It Is Assumed

Let's start with what you actually experience:

- You do not feel rotation.
- You do not feel orbit.
- You do not feel motion unless there's acceleration or change in direction.

You only perceive motion when there is:

- Resistance
- Friction
- Force
- Disruption of equilibrium

So if Earth were really spinning and orbiting at such extreme speeds, wouldn't you feel at least some of it?

Mainstream science says:

"You don't feel it because everything around you — the atmosphere, buildings, oceans — is spinning with you. It's all relative. You're in a perfectly smooth, inertial system."

But this answer is a philosophical patch. It avoids the deeper contradiction:

If everything is spinning, then what defines stillness?

Because if you feel still — then either your senses are lying, or the model is wrong.

What Is a Reference Frame, Really?

In Newtonian and Einsteinian physics, a reference frame is a perspective from which you measure motion.

If you're on a train moving at constant speed, and you throw a ball straight up, it comes right back down — because, from your frame, the ball hasn't shifted horizontally.

But if someone is watching from outside the train, the ball arcs forward.

So whose version is "correct"? According to mainstream physics: both are valid. Motion is relative to the observer's frame.

But UED questions this entire notion.

What if all these frames are just perceptual constructs, not actual physical truths?

What if your sense of stillness is not an illusion, but a reflection of your resonance with the field?

The UED View: Stillness Is the Default, Not Motion

In Universal Energy Dynamics, the universe is not made of moving objects, but of structured stillness.

What we call "motion" is a ripple consequence — the resolution of imbalance through the Structured Vacuum Energy (SVE) field.

You, your body, your awareness — are ripple participants embedded in that field.

When you feel still, it's not because you are being passively carried by hidden forces. It's because your ripple state is in complete alignment with the surrounding field.

There is no *relative* stillness — there is true resonance.

This means:

- You don't feel spin because there is none.

- You feel still because you are.
- Everything else — orbits, rotation, galactic drift — is a myth built on reference assumptions, not on felt reality.

The Illusion of Reference Shift

The modern heliocentric worldview depends on reference switching:

- Earth is spinning, but you treat it as still when calculating local motion.
- Earth is moving through space, but you draw maps as if it's stationary.
- Earth is an observer, but you plot other planetary paths as if it were fixed.

This dualism allows equations to work — but at the cost of truth consistency.

Let's expose the contradiction:

If you say the Earth spins, but use Earth as a still frame for every physics experiment, then you're not honoring motion — you're hiding it.

You're saying, "The Earth is moving, but let's pretend it isn't — so our numbers make sense."

That's not science. That's mathematical puppetry.

Ripple Alignment vs Physical Motion

Under UED, your experience of stillness is not a psychological trick.

It is a perfectly natural result of ripple field coherence.

Here's how it works:

- You (and everything around you) are embedded in a stable ripple basin — a harmonic field.
- If no external ripple disturbances occur, your body feels zero net ripple change.
- That condition is stillness — not relative, but absolute in your field context.

So the reason you feel no spin is simple:

You aren't spinning. You're not in motion. You're in stable ripple lock with the Earth field.

Airplanes, Birds, and the Myth of Moving Earth

Let's consider airplanes.

If Earth is spinning at 1,674 km/h to the east, then:

- A westward flight should be significantly shorter (the destination should rotate *toward* the plane).
- A hovering helicopter should see the ground move beneath it.
- A balloon should drift as the Earth rotates away.

But this doesn't happen.

Why?

Conventional answer: "The atmosphere moves with the Earth, so everything's synchronized."

UED answer:

There is no motion. There is only relative ripple stability.

The airplane flies because it breaks local equilibrium. But Earth's field remains ripple-locked. There is no rotating base frame to catch up with — because there is no base in motion.

Everything moves *within* the field — but the field itself is still.

The Stillness You Ignore

Stillness isn't just a lack of movement. It's an affirmation of alignment.

In a world dominated by motion-based thinking, stillness becomes suspicious.

But ask yourself:

- Why does calm feel peaceful?
- Why do you seek silence, balance, pause?
- Why do you intuitively trust stability over speed?

Because your body — your field — recognizes stillness as truth.

You don't feel spin because it doesn't exist.

What you feel is honest field alignment, betrayed by the story of motion.

Final Thought for This Chapter

They told you that you're spinning at high speed — and you believed them, even though you never felt a thing.

Now you know why.

Stillness is not absence of data — it is the presence of full field coherence. The body knows. The senses know. Only the story lied.

The Earth doesn't spin. You don't feel motion because you are in perfect harmony with the field of stillness.

It's not what you're missing — it's what you've forgotten.

11: Foucault, Gyroscopes, and the Misread Proofs

Field response vs mechanistic explanation

Ask any physicist or science student: "What is the proof that the Earth rotates?" You'll likely hear two names — uttered with pride and confidence:

- Léon Foucault
- The Gyroscope

These are presented as mechanical confirmations that the Earth spins. They're showcased in museums, classrooms, textbooks, and scientific demonstrations.

But what if these so-called "proofs" don't actually prove what we've been told? What if they are ripple responses, misinterpreted through a mechanical lens?

This chapter reexamines two of the most iconic "evidences" of Earth's rotation and reframes them under Universal Energy Dynamics (UED). What appears to be spin confirmation is, in truth, a field's reactive behavior, not a global mechanical motion.

The Foucault Pendulum: The Myth of Precession

In 1851, Léon Foucault suspended a 28-kg iron ball on a 67-meter wire from the dome of the Panthéon in Paris.

As the pendulum swung back and forth, its plane of oscillation appeared to rotate slowly over time. The explanation was revolutionary:

"The Earth is rotating beneath the pendulum."

The demonstration became legendary — and Foucault's pendulum became the most elegant and poetic proof of Earth's spin.

But let's ask some uncomfortable questions:

- What is actually moving — the Earth or the pendulum's field of alignment?
- Why does the rotation rate vary with latitude?
- Why are pendulum precession rates inconsistent in some locations or sensitive to environmental conditions?
- Why does the pendulum not drift off course instantly, but only shows slow angular drift?

In Newtonian or Einsteinian models, these questions are rarely explored in depth. But UED demands we go further.

The UED View: Ripple Lock and Field Drift

In Universal Energy Dynamics, the pendulum is not drifting due to Earth spinning underneath it. Instead, what you observe is:

A slow realignment of ripple coherence between the pendulum and its surrounding structured vacuum energy field (SVE).

Let's reframe it:

- The pendulum is a ripple-anchored mass, suspended in local field equilibrium.
- As it swings, it tries to maintain its ripple alignment with the still field.
- However, as Earth's layered field alignment changes minutely over time — due to local interference, cosmic field rhythms, or even solar-magnetic field interaction — the plane of ripple response subtly shifts.
- This slow re-alignment creates the illusion that the Earth is spinning beneath it.

But it is not the Earth rotating. It is the pendulum's tether to the structured field slightly realigning over time — much like a compass needle adjusting in a changing magnetic gradient.

So what we call "precession" is not mechanical — it is a field recalibration effect.

The Latitude Argument Debunked

Supporters of the Foucault pendulum claim that its precession proves Earth's rotation, citing this pattern:

- At the poles, the pendulum's plane appears to rotate once every 24 hours.
- At the equator, there is no rotation.
- At other latitudes, the rate varies by the sine of the latitude.

Sounds scientific, right?

But here's what's being ignored:

This same behavior is exactly what we would expect if ripple alignment fields radiate outward from a polar axis of stability and decay toward the equator.

Why?

- Because Earth's polar field tension is more focused, allowing stable ripple orientation.
- Near the equator, ripple alignment is more diffuse, unstable, and dynamically shifting — hence, the pendulum doesn't precess.

So yes — the latitude-dependent behavior exists. But it does not prove spin. It proves field variation across Earth's structure, which is expected in a UED framework, where ripple pressure and density vary with position and cosmic exposure.

The Gyroscope: Stability or Spin Illusion?

Gyroscopes — devices that maintain orientation due to angular momentum — are another popular "proof."

In airplanes, submarines, and satellites, gyroscopes are used to track movement. If the Earth spins, we are told, a stationary gyroscope should drift slowly to reflect that movement.

And indeed — some gyroscopes do show this drift. But UED asks again:

What is actually being measured — a drift due to Earth's motion, or a ripple-phase reorientation in a changing field?

Here's what UED proposes:

- A gyroscope is not an object "resisting" change because of mechanical inertia.
- It is a ripple-stabilized oscillator interacting with the

surrounding field.
- Any shift it exhibits is a sign of field tension gradient, not mechanical drift due to planetary rotation.

Gyroscope precession, then, becomes:
A field sensitivity effect, not a motion detector.
This explains:

- Why some gyroscopes show measurable drift and others don't.
- Why fiber-optic gyroscopes (based on light interference) sometimes detect anomalies that conflict with spin models.
- Why drift rates are often inconsistent or require post-correction and filtering.

If Earth were truly spinning smoothly and uniformly, such devices should show perfect, measurable rotation consistently everywhere. But they don't. Why?

Because there's no global spin. There's only local field noise — ripple recalibration.

The Hidden Circular Reasoning

Here's the biggest issue:

Foucault pendulums and gyroscopes are considered proof of spin because they behave as expected within a spin-based model.

But that's circular logic:

- You assume Earth spins.
- You interpret pendulum drift as confirmation.

- You declare the assumption validated.

But what if the same behavior can be explained by ripple realignment in a still Earth model?

Then the proof collapses.

Reframing the Apparatus: From Mechanics to Field Participants

Let's draw the distinction:

Apparatus	Traditional Interpretation	UED Interpretation
Foucault Pendulum	Rotational reference drift	Ripple field recalibration
Mechanical Gyroscope	Angular momentum resisting motion	Ripple-locked spin state adjusting to field pressure
Optical Gyroscope	Light-phase shift due to movement	Phase interference due to ATT changes in local field

In all three cases, mechanics are only the medium. The real player is the structured field. The behavior is a ripple response — not inertial confirmation of a rotating Earth.

Final Thought for This Chapter

The greatest illusion is not the spinning Earth —It's the false certainty of our instruments.

We built devices based on mechanical logic, then interpreted their ripple behavior as mechanical confirmation.

But the pendulum isn't dancing to Earth's rotation. The gyroscope isn't whispering secrets of planetary spin. They're both responding to ripple tensions, adjusting quietly as the field shifts around them.

Stillness is not broken by drift. It's revealed in how systems seek to re-align.

There was never a spin to prove — only ripples being misread.

12: Satellites and the Spaceframe Illusion

How ripple equilibrium can hold orbit without revolution

Among the strongest symbols of Earth's revolution narrative are satellites. We are told they orbit the Earth due to a balance of inertia and gravity — falling around the Earth perpetually, like celestial yo-yos. Their presence is often cited as final proof that:

"The Earth is moving, and satellites confirm it."

From communication relays to GPS systems, satellites supposedly track Earth's spin, match its rotation (as in geostationary satellites), or orbit in different planes to cover the globe.

But here's the fallacy buried under this narrative:

- We observe satellites moving across the sky.
- We receive data from them.
- We calculate trajectories using Newtonian or relativistic dynamics.

And from these observable consequences, we assume a rotating Earth with orbiting satellites.

But what if those calculations only *work* because of assumed motion? What if satellite behavior doesn't require revolution at all, but instead reflects a deeper field interaction?

In Universal Energy Dynamics (UED), we propose a radical reframing:

Satellites do not orbit because they are in freefall around a moving planet. They are suspended in ripple equilibrium zones within the Earth's structured vacuum field.

There is no revolution. There is ripple lock.

The Classical Model: Orbit as Perpetual Free Fall

Let's first revisit the conventional story:

- A satellite is launched tangentially at high speed.
- Gravity continuously pulls it toward Earth.
- Because of its tangential velocity, it keeps falling *around* the Earth — never hitting the surface.
- This "balance" of forward motion and downward pull creates a stable orbit.

In this model:

- The satellite is always moving.
- Earth is spinning or orbiting beneath.
- The entire system is a giant cosmic machine governed by force, velocity, and curvature.

But this model depends entirely on two assumptions:

1. That Earth is in motion (spin + orbit).
2. That satellites are following Newtonian paths, constantly "falling" in the geometry of space.

What happens if those assumptions are wrong?

Let's find out.

UED View: Ripple Zones, Not Orbits

In Universal Energy Dynamics, the universe is a structured field of stillness, where all visible phenomena are ripple consequences — not object-based motions.

This includes satellites.

Here's the reframed model:

The Earth is still — embedded in a stable ripple basin within the Structured Vacuum Energy (SVE) field. Around Earth, ripple pressure zones exist — areas where condensed ripple activity achieves local equilibrium. When a satellite is placed in such a zone with proper ripple momentum, it enters a state of ripple lock — where it appears to remain suspended or to "orbit."

There is no falling around Earth. There is no forward velocity vs gravity balance.

There is field participation — a dynamic, self-stabilizing interaction between the satellite's ripple structure and the Earth's ripple geometry.

What Keeps a Satellite "Up"?

In the Newtonian model:

- Gravity pulls down.
- Momentum pushes forward.
- The result is a curved path (orbit).

In UED:

- The satellite is not "held up" or "in motion."

- It is resting in a high-tension ripple node, where gravitational ripple pressure balances the satellite's own ripple condensation.

Just like a cork floating in a whirlpool — not moving around the center, but held in dynamic equilibrium — a satellite stays suspended not because it moves, but because it remains aligned with a ripple tension band.

If the field tension changes, the satellite:

- Drops in altitude.
- Drifts into a decay spiral.
- Appears to de-orbit.

But these are not results of motion. They are consequences of ripple disalignment.

Geostationary Satellites: A Ripple Illusion

Let's talk about geostationary satellites.

These are satellites that appear to hover over one fixed point on Earth, matching Earth's supposed rotation.

Conventional explanation:

"The satellite orbits Earth once every 24 hours, so from the ground, it appears stationary."

But that means:

- Earth must rotate.
- The satellite must move at exactly the right speed.
- Both systems must perfectly sync, indefinitely.

What UED says instead:

Geostationary satellites are ripple-locked objects placed in a field resonance band where the Earth's structured field maintains coherent signal alignment with the satellite's position.

There is no need for spin matching. There is no need for revolution. Just persistent ripple stabilization — like a calm bubble in a whirlpool of tension.

They appear fixed, not because of mechanical velocity balance, but because of equilibrium in perception-triggering ripple gradients.

Why Satellites Don't Fly Away or Fall

In a motion-based model, satellites must fight gravity by moving "fast enough" sideways.

In UED:

- Satellites don't "fight" gravity.
- They harmonize with ripple field pressure.
- When aligned, the ripple field neutralizes gravitational tension, creating a "float state" within the field.

You can think of it like a ripple hammock — held up by tension from all sides, even in a seemingly empty zone.

This view explains:

- Why satellites maintain orbits for decades.
- Why orbital corrections are subtle nudges, not constant thrusts.
- Why space stations "stay up" with little continuous effort.

They're not in motion. They're resting on invisible field scaffolding — and when that scaffolding shifts, they shift.

What About Satellite Trajectories and Velocity?

Critics might say:

"But we *measure* satellite velocities! We see them move across the sky! We calculate their orbital mechanics!"

UED doesn't deny that apparent motion is observed. But it reinterprets what is actually happening:

The observed "motion" is not the satellite changing position through space — It is the ripple field reactivating visibility zones sequentially, giving the illusion of movement, just like how the Moon phases are not about its rotation, but ripple reentry into perception.

Satellite "movement" is the field's shifting relationship with the observer, not the satellite's locomotion through curved space.

This also explains:

- Why satellites often disappear into Earth's shadow (not because of position, but due to visibility threshold loss).
- Why some satellites remain visible for long durations — they're not moving faster, they're held longer in field resonance.

Orbit Is Not Motion — It's a Field State

Let's summarize the contrast:

View	Conventional Model	UED Model
Satellite	Falling object in curved path	Ripple-locked structure in field
Orbit	Balance of velocity and gravity	Stable field resonance zone
Gravity	Force pulling inward	Ripple pressure pushing into tension
Geostationary	Matches Earth's spin rate	Static alignment in field
Velocity	Required to maintain orbit	Emergent appearance due to ripple realignment

This reframing doesn't deny the data — It reclaims the meaning of the data.

Final Thought for This Chapter

Satellites are not racing marbles around a spinning sphere. They are suspended signatures — ripple knots held delicately in a field we never learned to see. They whisper truth from stillness: "I'm not moving. I'm aligned."

The Earth doesn't move. Satellites don't orbit. They all simply coexist in ripple equilibrium — And what you see as revolution...Is just the universe resolving itself into coherence.

Interlude: When the Ground Refuses to Move

Twelve chapters in, and the ground beneath your mental feet no longer spins.

You've stepped outside the revolving classroom model. You've stared directly at the Sun without assuming it's static. You've felt the stillness that's always been there—beneath stories of orbits, revolutions, and forces.

You've questioned:

- Why day and night must imply rotation.
- Whether elliptical revolutions are evidence or interpretation.
- Why gravity is felt but never explained.
- Why you feel still even when they say you shouldn't.
- And whether instruments like pendulums and gyroscopes are confirming spin—or simply dancing with a field we don't yet understand.

And finally, you've seen that satellites — the supposed crown jewels of Earth's orbital truth — may simply be ripple-bound passengers, not free-falling orbiters.

So what now?

We've dismantled the myth of motion. Not by denying science, but by taking it seriously — all the way back to stillness. Not by mocking the models, but by asking if they mistook **appearance for cause**.

And in doing so, we've exposed a quiet truth:

What we call "motion" is often **the field reconfiguring**, not an object moving. What we call "spin" is a **reference shift**, not a rotation. What we call "orbit" is a **consequence of resonance**, not a path of revolution.

It turns out the Earth never needed to move to explain the sky.

So now, as we move into **Part III**, we begin to rewrite — from scratch — The very definitions of **motion**, **inertia**, **acceleration**, **momentum**, **light**, and **energy**.

This is not about standing still. This is about **re-centering truth** around stillness.

Because once you've unspun the Earth, everything else needs to be untangled too.

Let's rebuild the laws — not for tradition's sake, but for truth's.

Part III – Rewriting Physics

13: Redefining Inertia

Ripple Lock in Still Fields

Your earlier insight fully expanded

"Inertia" is one of the first concepts we learn in physics. We're told it explains why things don't change motion easily. Why an object in motion stays in motion, and why an object at rest stays at rest—unless acted upon by an external force.

But here's the problem:

We memorized the behavior, but we never understood the reason.

In Newton's formulation, inertia is a given—an intrinsic property of mass. No explanation is offered for where it comes from or why it exists.

We're told: "That's just how matter works."

But the deeper we look, the stranger it becomes:

- What actually keeps an object moving when nothing is pushing it?
- Why should something resist motion change?
- Why does inertia vary with mass—and what even *is* mass?

- Why is this resistance *uniform* in space?

The answers don't exist in the classical framework—they're brushed off as "natural laws." But UED doesn't settle for assumptions.

In Universal Energy Dynamics, inertia is not a property. It is not built-in. It is not fundamental.

Instead, inertia is ripple-lock—a dynamic state of equilibrium between a ripple-structured object and the still field it's embedded in.

Inertia is not a thing an object has—it's what an object does when its ripple alignment is undisturbed.

The Classical Definition of Inertia (and Its Gaps)

In classical mechanics:

- Inertia is defined as "the tendency of an object to resist changes in its state of motion."
- It's linked directly to mass: the more mass an object has, the more inertia it exhibits.
- Newton's first law *defines* this behavior as a principle—not as a consequence.

But notice what's missing:

- *Why* does the object tend to resist change?
- What enforces this resistance?
- What sustains the motion without energy input?
- What prevents the natural decay of movement in a frictionless universe?

Physics never explains. It simply *describes*. And over time, description hardened into doctrine.

The UED View: Inertia as Ripple Lock

Let's strip the illusion.

In UED, we don't see "objects" as isolated particles moving through space.

Instead:

- Every object is a localized ripple structure — a condensed energy pattern.
- These patterns are locked into a surrounding field of stillness — the Structured Vacuum Energy (SVE).
- When a ripple structure aligns with the surrounding field and faces no resistance, it stays in that alignment without change.

This stable ripple-state is perceived as:

- An object staying at rest, or
- An object continuing at constant velocity.

But there is no velocity in itself. There is only persistent ripple orientation.

Inertia is the field's confirmation that no disruptive ripple pressure exists — the ripple remains locked in its configuration.

Why This Changes Everything

If inertia is ripple-lock, not property, then:

- Inertia is not intrinsic—it is contextual.
- It depends not on the object alone, but on its field participation.

- If the field shifts, inertia dissolves—even if mass stays constant.
- Inertia can appear or vanish as a response to ripple disturbance.

This explains phenomena such as:

- Sudden energy loss in collisions (ripple decoherence).
- Why inertia feels "massive" in gravitational zones (ripple compression increases field tension).
- Why inertia appears absent in microgravity (ripple field is diluted, not nullified).

In other words:

Inertia is not resistance to change — it is the absence of ripple contradiction.

A Thought Experiment: The Ice Skater and the Ripple Field

Imagine a skater gliding effortlessly across a perfectly smooth, frictionless ice surface.

In classical terms:

- The skater keeps moving due to inertia.

In UED terms:

- The skater's motion is a ripple alignment established through the initial push.
- Once in ripple lock, the body and field are in dynamic agreement.
- No ripple disturbance → no change → continued ripple

expression → motion.

But now imagine:

- A small ripple disturbance (wind, vibration) disrupts the alignment.
- The skater's motion deviates or slows—not because of friction, but because ripple lock has broken.

What we called "inertia" was never internal—it was external field agreement.

Ripple Lock vs Force: A Shift in Perspective

In Newtonian mechanics:

- Force breaks inertia — it "overcomes" the object's tendency to resist.

In UED:

- There's nothing to "overcome."
- Force is not acting *against* inertia.
- Force is the field's signal that ripple alignment has changed.
- When ripple lock breaks, the object reorients to a new field dynamic—and we interpret this as "acceleration."

So the real question is never "what force acts?" It's "what changed in the ripple coherence?"

The Deep Implication: Rest and Motion Are the Same

In UED, rest and constant motion are not different states. They are both stable ripple locks.

- When ripple symmetry holds → stillness or straight-line motion.
- When ripple symmetry breaks → change, acceleration, force.

Therefore:

Motion isn't movement through space — it's duration of ripple coherence.

This resolves the greatest paradox in Newtonian physics:

How can something keep moving forever without being powered?

Answer: Because it's not moving. It's simply not disturbed.

Inertia and Consciousness

Let's go even deeper.

In biological systems, inertia appears as habit—the tendency to keep acting, thinking, or perceiving in a certain way.

In UED, this is not metaphor. It's the same mechanism.

- Thoughts are ripple structures.
- Behaviors are ripple feedback loops.
- Habits are mental ripple locks.
- Resistance to change is psychological inertia.

When ripple fields in the brain align in a stable feedback state, you continue with the same behavior—even if it no longer serves you.

This is why:

- Inertia is not just physical.

- It is a universal ripple condition—across matter, energy, and mind.

Final Thought for This Chapter

Inertia was never inside the object. It lived in the space between—in the field, in the stillness, in the silence that cradled motion.

It is not a trait. It is a truce. A contract between energy and field: *"I won't change if you don't disturb me."*

Inertia is not the resistance to change. It is the field's silent agreement to hold form—until it doesn't.

Stillness is not a lack of energy. It is the perfect agreement of energy with its ripple space.

That's inertia.

And from this foundation, we now move to rethink not only motion, but acceleration, momentum, and energy itself.

14: Acceleration

A Field Re-tuning, Not Just Vector Change

Why acceleration is a state conflict in ripple flow

In conventional physics, acceleration is simple. It's defined as: "The rate of change of velocity with respect to time."

In formula: $a = \Delta v / \Delta t$

It tells us how fast something is speeding up, slowing down, or changing direction.

But behind this neat mathematical expression lies a deeper assumption — That objects possess motion inherently, and acceleration is just an external input that alters that motion.

But what if this idea is fundamentally wrong?

What if acceleration is not a mechanical change in speed or direction, but rather a conflict in the ripple field — a re-tuning process where ripple alignment is disrupted and seeks a new balance?

This is what Universal Energy Dynamics (UED) reveals.

Acceleration is not a push or pull. It is a symptom of ripple misalignment — a phase conflict within the structured field. The observed acceleration is just the body's response to ripple field tension trying to find a new harmonic state.

The Classical Illusion: Acceleration as Applied Force

In Newtonian physics:

- When you apply a force to an object, it accelerates.
- The mass resists change (inertia).
- The force "overcomes" this resistance and causes velocity to increase or decrease.

This model works well for calculation. But it doesn't tell us what acceleration *is*. It doesn't explain why an object speeds up beyond "because we applied force."

It's a descriptive layer, not a causal one.

Acceleration is treated as a mechanical event, but never as a field-based phenomenon.

The UED View: Acceleration as Ripple Re-tuning

In Universal Energy Dynamics, acceleration is reframed:

Acceleration is the field's response to a ripple misalignment caused by external disturbance or internal field change. It is not motion change — it is field reconfiguration.

Here's how it works:

1. An object is in ripple lock with its surrounding field — either at rest or in constant motion.
2. A new ripple influence (from another energy input, interaction, or force) disturbs that lock.
3. The object's ripple structure now faces internal tension — it is out of tune with the local field.
4. This tension creates conflict — a need to shift to a new equilibrium state.
5. The process of shifting from one ripple alignment to

another is what we perceive as acceleration.

The acceleration ends when a new stable alignment is achieved.

Acceleration ≠ Motion Increase

In UED, we don't define motion as a thing that "increases." Instead:

- Constant motion = ripple coherence held steady
- Acceleration = ripple misalignment forcing a new pattern
- Deceleration = ripple tension decreasing toward stillness

This means:

- You don't speed up — your field reorients.
- You don't slow down — your ripple tension decreases.

Motion never "builds up." It shifts, re-grounds, or disperses.

So acceleration is not a curve on a graph — It is a distortion zone in the field, seeking balance.

A Practical Example: Car Acceleration

Let's use a simple analogy:

A car speeds up on a highway.

In classical terms:

- The engine applies force through wheels.
- Friction with the road generates forward motion.
- Velocity increases, acceleration is calculated.

In UED:

- The car is not "pushed" forward.
- The engine's energy modifies the ripple state of the car's body-field.
- The surrounding field now resists this new ripple flow, creating internal tension.
- The car-body's ripple structure begins reorienting itself, step by step — that's the acceleration.
- As it aligns to a new ripple flow, the system returns to local coherence — and acceleration ceases.

You're not increasing motion. You are re-tuning your place in the ripple field.

Why Acceleration Feels Heavier than Constant Velocity

Have you noticed this?

- You don't feel motion at constant speed.
- But you feel acceleration — in your chest, in your seat, in your gut.

Why?

Because acceleration is ripple tension building inside your body.

- Constant motion = ripple lock = no tension.
- Acceleration = ripple conflict = pressure, sensation, distortion.

Your body reacts to ripple dissonance. You "feel" acceleration not because you're changing speed, but because your field is trying to adapt under tension.

This is why astronauts feel massive g-forces at launch — not because they're moving fast, but because they are in intense ripple realignment under collapsing field equilibrium.

Acceleration and Consciousness

Like inertia, acceleration has a psychological parallel.

Think of:

- Emotional change
- Behavioral transformation
- Sudden shifts in attention or awareness

In each case, what we call "mental acceleration" is conflict in the conscious field.

You experience:

- Discomfort
- Strain
- Uncertainty

Why?

Because the mind is re-tuning its internal ripple alignment — shifting from one stable pattern to another.

In both matter and mind:

Acceleration = friction of change, not gain of momentum.

It's not a gain. It's a field negotiation.

Redefining the Equations

Let's reframe the classical equation:

$a = \Delta v / \Delta t$ = change in velocity over time
In UED terms:
$a \propto \Delta R / ATT$ Where:

- ΔR = change in ripple alignment
- ATT = alignment threshold time — the time it takes for ripple coherence to reach perceptual activation

So acceleration becomes:

- Not a rate of motion change,
- But a rate of ripple re-alignment within the field

It's the energetic "lag" of coherence catching up with intent.

Acceleration Ends Where Harmony Begins

Acceleration is not sustainable. It always ends — not because motion completes, but because:

Ripple conflict resolves, and a new alignment state is achieved.

This is why:

- Objects reach terminal velocity.
- Systems return to rest.
- Bodies adapt to change.
- Stress fades once transformation completes.

What you call "end of acceleration" is the moment of field peace — where ripples stabilize once more.

Final Thought for This Chapter

Acceleration was never speed.

It was your ripple body arguing with the field — and eventually coming to terms with it.

It's not a mechanical climb. It's a field negotiation between what is and what must become.

In motion, we glide. In rest, we settle. In acceleration, we evolve.

That's not physics. That's ripple psychology — embedded in the structure of stillness.

And with this, we now prepare to explore momentum, not as a vector quantity, but as stored ripple intent waiting to express consequence.

15: Motion Redefined

From Path to Participation

The Ripple View of Object Movement

In traditional physics, motion is treated as simple, linear, and measurable. It's defined as:

"A change in position of an object with respect to time."

We draw graphs of distance vs. time. We assign coordinates, directions, and speeds. We track objects on paths, and claim to understand motion as long as we know where something was, where it is now, and how fast it's moving.

But if you pause and reflect — *what is motion really?*

- Is it a change in location?
- A consequence of force?
- A result of time progression?

The deeper you go, the less convincing the textbook answer becomes.

Physics measures motion. But it doesn't know why it happens. Or what actually moves in the first place.

UED (Universal Energy Dynamics) says: Motion is not a "thing" or a "path." Motion is a field participation — a ripple expression born out of alignment shifts in structured stillness.

The Classical Illusion: Motion as a Traceable Line

Under Newtonian and Einsteinian physics:

- Objects are *things* that move *through* space.
- Their motion can be described by position over time.
- Forces initiate or alter motion.
- Trajectories are predictable — unless disturbed.

It's all very clean — on paper.
But observe the real world:

- You don't "feel" the Earth move, even though it allegedly spins at 1,600 km/h and orbits the Sun at 108,000 km/h.
- You don't "see" electrons moving in circular orbits, even though they're said to be in constant motion.
- You don't "experience" the expansion of space, even though galaxies recede at nearly light speed.

So what are we really perceiving?

Are objects tracing paths, or are we interpreting shifting ripples as motion?

The UED Lens: Motion as Ripple Consequence

In UED, motion is not an object's travel through emptiness.

Motion is the observable effect of ripple alignment shifting within the structured field of stillness (SVE).

In other words:

- When an object "moves," it's not *changing position* in a passive void.
- It is engaging with the surrounding ripple field, re-participating in a new configuration.
- The observer perceives the transition from one ripple equilibrium to another — and labels this as motion.

So:

- No ripples = no motion
- Stable ripple lock = apparent stillness or constant motion
- Changing ripple alignment = what we call acceleration, deviation, or interaction

This means motion is not fundamental — it is always a consequence of ripple behavior.

A Ripple-Based Breakdown of Motion

Let's look at how motion arises, step by step:

1. **Stillness:** The object exists in a stable ripple configuration. It appears "at rest."
2. **Ripple Disturbance:** An energy input — external or internal — perturbs the field equilibrium.
3. **Ripple Reconfiguration:** The object responds to this imbalance, its ripple field begins realigning.
4. **Perceived Change:** The observer perceives the shift in alignment as a change in position over time — i.e., motion.

So motion is not about going somewhere. It is about becoming somewhere else — through field resonance.

The body didn't travel. The ripples reconfigured, and the body participated in a new position.

Thought Experiment: The River vs the Ripple

In mainstream physics, motion is like watching a leaf float down a river.

In UED, the leaf is not moving through water. Instead, the river itself is a standing ripple field, and the leaf's position is a point of constructive alignment within that ripple.

When the ripples shift — either from inside (the leaf's own field) or outside (another ripple disturbance) — the position of harmony changes.

We then perceive the leaf as having "moved." But in truth, the field changed — not the "leaf" in isolation.

This means:

The world is not a stage across which actors move. The world is a ripple arena where participation patterns evolve.

The Map Deceives: Motion Is Not About Position

Consider this analogy:

- A train is said to move from point A to point B.
- On a map, this looks like a line.

But what truly happens?

- The train is a massive ripple-bound structure.
- Every micro-alignment between its engine, tracks, atmosphere, Earth's crust, and field of gravity must update *in sync* for "motion" to occur.

- These are field consequences, not isolated displacements.

So motion isn't about the train changing coordinates. It's about field states changing across layers of ripple interaction.

The map simplifies. Reality ripples.

No Motion Without Observer Alignment

Here's where it gets deeper:

Motion doesn't exist without a participating observer.

Why?

Because motion is only perceptible when the observer's field alignment differs from that of the moving object.

If both are ripple-locked together in the same flow → no motion is perceived.

This explains:

- Why you don't feel motion in a smooth aircraft.
- Why the Earth seems still despite orbital speeds.
- Why people in dreams or imagination seem to move — but actually, only your mind's field is realigning.

There is no absolute motion. There is only differential ripple participation between fields.

Motion and Consciousness: Beyond the Physical

Let's look at mental and emotional "motion":

- A change in thought
- A shift in mood
- A movement of intent

These are not paths traced in the brain. They are ripple transitions in the conscious field.

You "move" emotionally when your ripple alignment shifts — not when a neuron fires.

You "move" through memories, not in time, but by re-entering ripple patterns stored as field potentials.

So:

Thought is motion. But not because anything travels — because awareness participates differently.

Final Realization: Motion Is Meaningful Ripple Reentry

Motion, in its truest form, is:

- Re-participation in a new ripple alignment
- A shift in resonance across field boundaries
- A re-expression of the same stillness under new configuration

So we now reject:

- The idea of "empty space" through which things move
- The assumption that motion is an object's isolated property
- The belief that time and position define movement

And we accept:

Stillness is the canvas. Ripples paint the change. Motion is the eye watching the pattern shift.

You never really moved. You just kept waking up in different versions of the same stillness — because the field allowed it.

Closing Statement for Chapter 15

Motion isn't how you go. It's how the universe lets you reappear — again and again — by re-aligning with the ripple field.

What we call "movement" is just permission to participate... one alignment at a time.

So next time you see a falling leaf, a walking child, or a comet across the sky, remember:

Nothing is traveling. Everything is participating.

16: Energy

The Intent of Ripple Displacement

No Movement, Only Structured Field Interaction

In modern physics, energy is the holy grail — the measurable essence of work, change, and existence.

From Newtonian mechanics to Einstein's relativity, from chemical bonds to nuclear fusion, energy has been used to explain everything:

- If something changes → energy was involved.
- If something moves → energy was transferred.
- If something transforms → energy was conserved.

But what is energy?

Textbooks define it as:

"The capacity to do work."

That's a circular definition. It replaces cause with consequence — work defines energy, energy causes work.

So again: What is energy? Where is it stored? What truly moves? And how is it transferred — if, in reality, nothing actually moves at all?

UED — Universal Energy Dynamics — answers it simply:

Energy is not something objects carry. Energy is the *intent* expressed through ripple displacement in structured stillness.

In short:

There is no movement. Only structured field interaction — and what we call "energy" is the degree and direction of participation in that field.

Energy as Intent, Not Substance

Energy is usually imagined like a fluid:

- Stored in a battery
- Carried in a moving mass
- Released in heat, light, or motion

But this view is mechanistic — it sees energy as stuff passed around.

UED radically redefines this:

Energy is a ripple tension — a directional intent to disturb, balance, or reconfigure the field.

It's not a quantity flowing in space. It is a structured will — a programmed imbalance expressed as ripple realignment.

So, energy is not:

- A fuel
- A wave
- A particle's property

It is:

- A potential consequence seeking resolution through ripple re-participation.

In other words:

Energy is not what moves things. It is what *urges* the field to allow apparent movement.

Why Energy ≠ Motion

In physics, kinetic energy is defined as:

$K.E. = \frac{1}{2}mv^2$

This implies energy depends on mass and velocity — the more an object moves, the more energy it carries.

But UED shows:

- Mass is not intrinsic — it's ripple condensation.
- Motion is not real — it's ripple alignment.

Therefore, kinetic energy is not real movement — it's just the field's active intent to continue a ripple pattern.

There is no "energy of motion." There is only tension in ripple alignment.

And when a system is "at rest," it's not that energy is gone — it has simply returned to balanced participation, zero intent.

So:

- No one carries energy.
- The field holds the potential.
- The object is just the channel of alignment.

Energy Transfer: A Reinterpretation

Conventional physics talks about:

- Heat transfer
- Electric current
- Energy conversion

- Work done

But if no object is moving in empty space, how does energy "travel"?

In UED:

There is no transfer. There is only ripple reconfiguration, where localized field states influence one another — and appear to share energy.

So when you:

- Heat a metal rod
- Charge a battery
- Stretch a spring

You are not pushing energy into a system. You are tuning its ripple equilibrium — disturbing the structured field to store intent.

Later, when the spring is released or the battery used, the field relaxes, resolving that tension through new configurations.

This is energy — not substance, but structured desire.

Energy as a Ripple Dialogue

Imagine this:

- You drop a stone in a still pond.
- Ripples expand.
- They reach a leaf floating nearby.
- The leaf moves.

Did energy "travel" from stone to leaf?

Conventional physics says: yes — via waves, force, displacement.

UED says:

The stone's entry disturbed the field. The ripple reconfigured the still pond. The leaf participated in the realignment.

Nothing was "sent." The field was asked to shift, and it responded.

This is how energy flows — not through space, but through consent of field participation.

Matter and Energy: Not Equivalent, but Aligned

Einstein gave us the famous equation:

$E = mc^2$

He said mass and energy are equivalent — you can convert one into the other.

But UED offers a deeper insight:

- Mass is ripple condensation.
- Energy is ripple intent.

They are not the same — but they are interlinked.

Matter stores ripple density — a frozen pattern of field coherence. Energy is the will to change that pattern — the readiness to evolve, disturb, or align.

So $E = mc^2$ is a mathematical window into a ripple reality:

The more intense the ripple structure (mass), the greater its potential to distort (energy) if released.

But it doesn't mean energy is a "thing" hiding inside matter.

Rather:

- Matter is the memory of ripples.
- Energy is the language of ripples.

Consciousness and Energy: Two Sides of Ripple Intent

UED goes even further.
It sees a direct link between awareness and energy:

- Conscious intent = ripple disturbance in the mental field
- Physical energy = ripple disturbance in the material field

In both cases:
Energy is the expression of desire in structured stillness.
This is why:

- Mental fatigue and physical fatigue mirror each other.
- Thoughts can create emotional energy.
- Willpower can "charge" the body.
- Rest restores alignment — not because it "gives energy," but because it reduces ripple dissonance.

So energy is not just physics. It is psychophysics — a field phenomenon of intention.

Final Redefinition: What Is Energy?

Let's summarize:

✗ Energy is not a substance.

✗ Energy is not motion.

✗ Energy is not stored inside particles.

✓ Energy is a directional ripple imbalance

✓ It is the structured will to shift within a still field

✓ It is consequence in waiting, potential in tension, intent seeking resolution

That's why energy can:

- Heat
- Move
- Transform
- Build
- Destroy

Because in all these cases, it's not the object acting — it's the field reconfiguring to allow action.

Energy is stillness made active. Energy is silence prepared to speak.

Closing Reflection

From now on, never ask: "Where is the energy coming from?"

Ask instead: "What intent is disturbing this field, and how is it trying to resolve?"

Because behind every explosion, every orbit, every heartbeat, and every thought...Is the same question:

What wants to become?

And energy is simply the ripple field's answer: "Let me show you."

17: Momentum

Stored Ripple Potential, Not Linear Push

Rewriting the True Meaning of Motion Memory

In classical physics, momentum is defined as:

Momentum (p) = mass × velocity

A neat and measurable property. The faster something moves, and the heavier it is, the more momentum it has. And unless acted upon by an external force, it keeps going — Newton's First Law.

Momentum becomes the prized jewel of motion:

- It gives stability to movement.
- It explains collisions, rebounds, transfers.
- It forms the foundation of conservation laws — both in mechanics and in quantum fields.

But pause.

Ask yourself: What is momentum actually storing?

What is it within the object that preserves motion — even in empty space?

- Where is this "push" located?

- How can a stone in space keep "moving" without any continued input?
- Why does conservation of momentum seem so real, even when the environment seems to do nothing?

UED steps in with a radically clear view:

Momentum is not a linear push. It is stored ripple potential — a residue of past alignment seeking resolution through continued participation.

Let's unbuild the myth and rebuild momentum from first ripple.

The Classical View: A Push That Never Ends

According to Newton:

- You push an object.
- It accelerates.
- Once no more forces act, it moves uniformly — forever.
- This motion is "stored" as momentum.

But this definition assumes:

1. A vacuum with no resistance.
2. A frame of reference that stays still.
3. A body that somehow remembers its velocity.

But none of these is real:

- The vacuum is never truly empty — fields ripple even in space.
- Reference frames are constructs — the Earth, the

galaxy, even the cosmos move.

- And an object doesn't "store" velocity inside itself — it's not a spinning top with a gyroscope.

So what actually persists?

What allows a bullet, a satellite, or a drifting rock to retain "motion"?

UED answers: the field does.

Ripple Memory: What Really Moves

Every action causes a disturbance in the field — the structured vacuum energy (SVE).

This disturbance is not destroyed — it remains embedded, like a wrinkle on water.

When an object is pushed:

- It aligns with a new ripple configuration.
- The surrounding field remembers this alignment.
- As long as nothing interferes, that configuration remains valid.

Thus:

The object doesn't keep moving because of a force it carries. It keeps moving because the field has not resolved the previous ripple pattern.

The motion persists not inside the object, but outside — in the unresolved ripple footprint.

Why It Looks Like Momentum

To an observer, this appears exactly like Newtonian momentum:

- A car keeps rolling.
- A comet keeps flying.
- A soccer ball slides on ice.

But in UED:

- The object is just re-aligning with its ripple environment.
- What we call "velocity" is a chain of ripple permissions, not an internal state.
- And "momentum" is the duration of unresolved ripple intent.

Therefore:

Momentum = Time-extended ripple alignment due to unresolved energetic cause.

It's not just mass × velocity — it is field inertia × ripple intent.

Ripple Lock vs Ripple Drag

Let's now distinguish:

- **Inertia**: Ripple lock (stable participation in the field).
- **Momentum**: Ripple drag (continued alignment from prior disruption).

Inertia = being still with the field. Momentum = being in motion within the ripple memory of the field.

When the motion seems to continue:

- It's because the field ripple remains aligned.
- Not because the object carries energy like a backpack.

This is why in space:

- A rock doesn't "coast" endlessly.
- It resolves its field relationship until some interaction breaks it.

So conservation of momentum is not a law of motion. It's a consequence of unresolved ripple equilibrium.

Impulse and Interaction: The Ripple Shock

In Newtonian mechanics:

Impulse = Change in momentum (Force × time)

But in UED:

- Impulse is a sudden ripple shock — a sharp, asymmetrical disturbance in the field.
- The field doesn't "absorb" force — it absorbs tension and rearranges ripple patterns.
- This creates a new momentum path, not by pushing, but by realigning participation.

So next time two billiard balls collide:

- It's not the balls exchanging energy.
- It's the field resolving conflicting ripple zones into a new configuration.

What we observe as:

- Equal and opposite reactions,
- Elastic or inelastic collisions,
- Recoil and rebound,

...are simply different ripple resonance resolutions.

Quantum Momentum: A Deeper Echo

Even in quantum mechanics:

- Electrons have momentum.
- Photons carry momentum.
- Wave-particle duality demands "momentum transfer."

But in UED:

The quantum world is not a probabilistic fog — it is ripple localization.

Momentum there is not an abstract value.

It is:

- A directional tension in the field,
- Held until resolution,
- Observed only when re-participation occurs.

So when an electron "jumps," or a photon is "absorbed," we are not witnessing tiny particles transferring motion.

We are observing the field collapsing its previous intent, and resolving it through ripple consequence.

Momentum of the Mind?

Let's stretch this further:

- Emotional momentum
- Habitual momentum
- Social momentum

All of these behave like physical motion:

- Hard to stop
- Resistant to change
- Continuing despite new inputs

In UED Psychology:

These too are stored ripple potentials — unresolved alignments within mental or social fields.

That's why:

- Emotions linger even after events pass.
- Patterns repeat even when logic changes.
- Systems persist in behavior long after reason is gone.

Because momentum is ripple memory, not logic.

Closing Realization

So what is momentum?

✗ Not a push

✗ Not a force stored

✗ Not a property of mass

✓ It is the duration of unresolved ripple participation,

✓ It is the field's permission to continue a previously caused consequence,

✓ It is the path of least resistance for ripple reentry, once intent is expressed.

In that sense:

Momentum is not how fast you move. Momentum is how long the field lets you move — without asking new questions.

And when the field does ask?

You stop.

Not because friction won. But because the ripple decided to rest.

18: Light Doesn't Travel – It Appears

Ripple Alignment and the Role of ATT (Alignment Threshold Time)

We've grown up believing that light travels — from the Sun to Earth, from a bulb to your eye, from stars across galaxies. We've been told it has a speed limit — a universal constant: c = 299,792,458 m/s

This belief is so deeply embedded that it forms the backbone of:

- Einstein's relativity,
- Our GPS systems,
- All notions of causality and simultaneity in physics.

We think of light as:

- A stream of particles (photons),
- A wave propagating through space,
- Or sometimes both (thanks to quantum weirdness).

But here's the radical challenge UED offers:

Light does not travel. It appears. And what you see as speed is not motion — but the time it takes for the field to align.

This isn't poetry. It's precision.

The False Frame of Light in Motion

Let's question the basics:

- If you light a torch in the vacuum of space, where does the light "go"?
- Does it really carry itself, particle by particle, through nothingness?
- Or does the field already exist, and the light simply activates the visible alignment?

UED asserts that:

Light is not an object or wave traveling through space. Light is the activation of alignment between emitter and observer across the ripple field.

In simpler terms:

Light appears when the ripple consequence resolves visibility.

It's not motion. It's participation.

Ripple Field and ATT: The Real Mechanism

According to UED (Universal Energy Dynamics), the universe is not empty space, but a structured field — the SVE (Structured Vacuum Energy).

When an event happens (e.g., a spark, star ignition, eye blinking), ripples are generated in the field.

These ripples do not move like bullets. They modulate the field — just like pressing one end of a tight cloth can produce alignment effects at a distance, even before the "bump" physically arrives.

Enter: ATT – Alignment Threshold Time

ATT is the time taken for the field to align sufficiently for a consequence (like light) to be perceptible.

It depends on:

- Ripple density of the field
- Observer's position in the field
- Nature of the ripple (wavelength, energy, phase)

So the "speed of light" is not how fast something is moving. It's how fast alignment is achieved between the emitter's disturbance and the observer's field coherence.

What "Seeing" Actually Means

When you "see" light:

- You're not catching a photon in your eye like a baseball.
- You're not tracing a wave's journey from point A to B.

Instead:

You are participating in the ripple realignment that the emitter has triggered.

You are not receiving light. You are becoming visible to the same ripple consequence.

Think of it like this:

- You strike a bell.
- Someone else touches the bell far away, and feels the ring.

They didn't get "sent" the sound. They felt the field vibration when their point became resonant.

So too with light.

The Double-Slit Revelation: Not Particle, Not Wave

In quantum physics, the double-slit experiment famously confused everyone:

- Light behaves like a wave when unobserved.
- Light behaves like a particle when observed.
- Interference patterns disappear when you watch.

UED cuts through the paradox:

The ripple field holds all potential paths. But visibility is a consequence of alignment, not trajectory.

The light never "went through both slits." It appeared wherever the ripple configuration aligned with observer's ATT threshold.

Once observation occurs:

- The field collapses the ambiguity.
- The ripple aligns only with what's being resolved.

So there's no mystery.

There's only ripple selection — and the illusion of travel.

Star Light and Cosmic Delusion

Let's go cosmic.

They say:

- Light from distant galaxies takes millions of years to reach us.
- So we're "looking into the past."

But under UED:

That light is not *traveling* toward you like a spaceship.

Rather:

- The ripple field across the universe is always ready.
- When a condition (like Earth's field + ATT alignment) is met,
- The star's ripple appears visible to us — *now*.

So "light-years" aren't distances light travels. They are field response thresholds — how long it takes for field tension to stabilize for perceptual visibility.

Which means:

You're not seeing the *past* — You're witnessing the present consequence of ripple resolution.

Why Shadows Are Not Caused by Light

Let's now flip another assumption.

We think shadows are caused because:

- Light travels in straight lines,
- An object blocks the path,
- So the region behind is dark.

But UED proposes:

Light doesn't "hit" and get blocked. Light appears only when the field aligns.

A shadow is not a light deficit. It is a visibility void — where field alignment is not allowed due to ripple obstruction.

So:

- No need to imagine rays being cut.

- Just see where the field can't vibrate into coherence.

Why You Can't See the Moon at Noon

Ever wondered why the Moon becomes invisible during most of the day — even when it's technically in the sky?

Conventional answer: it's washed out by sunlight.

UED answer:

The Moon does not "reflect" light. The field resolves the Moon's presence as visible only when the ripple alignment reaches ATT threshold.

In sunlight's field dominance:

- The ripple priority goes to the Sun.
- The Moon's ripple doesn't resolve — it gets suppressed.

So visibility is a field selection, not just a matter of brightness.

Final Redefinition: What Is Light?

Let's distill the essence:

✗ Light is not a moving particle.

✗ Light is not a vibrating wave.

✗ Light does not carry itself across vacuum.

✗ Light does not "know" where it's going.

✓ Light is the appearance of resolved ripple alignment.

✓ It manifests when emitter and observer are aligned through the structured field.

✓ Its "speed" is a field characteristic, not a universal motion.

In this view:

Light is not what travels. Light is what the field permits you to see — when the ripple is ready.

Closing Thought

So next time someone says: "Light takes 8 minutes to travel from the Sun to Earth,"

Tell them: "No, friend. It takes 8 minutes for the field to align and for your eye to become part of that ripple consequence."

The Sun didn't send light. You met it.

And that's how vision — and truth — really works.

19: Time and Motion Are Not Fundamental

Real Phenomena Are Perceptual Consequences

From ancient astronomy to modern relativity, time and motion have always been at the center of physical understanding. They are the twin pillars holding up our temples of mechanics, thermodynamics, relativity, quantum theory — everything.

We measure motion to calculate time, and we use time to understand motion. We treat them as the fundamental threads of reality.

But are they?

What if both time and motion are not real building blocks of the universe...but just consequences of perception?

Let's walk this radical insight slowly, ripple by ripple.

The Legacy of Time and Motion

Classical physics treats motion as a change of position with time:

$v = dx/dt$

Relativity stretches this further — time is woven into space, forming spacetime, and motion becomes a path through this four-dimensional continuum.

Quantum theory disrupts certainty — adding probability and fuzziness to both motion and time.

But all these frameworks — classical, relativistic, quantum — assume one thing:

That time and motion exist independently, and we experience them as consequences.

UED flips this entirely:

Time and motion are not pre-existing realities. They are apparent outcomes of ripple interaction within a still, structured field.

In other words:

- We perceive time.
- We interpret motion.
- But neither is fundamental.

Stillness: The Real Background

In UED, the entire universe is built on a structured vacuum energy field (SVE).

This field is inherently still. No movement. No ticking. No unfolding.

Everything that we call:

- Motion (a ball flying, a planet orbiting),
- Time (a clock ticking, a life aging),

...emerges only when ripples disturb this field and perception resolves the disturbance into sequential experience.

Without ripples — there is no motion. Without perception — there is no time.

Ripple Interference Creates the Illusion of Movement

What we interpret as motion is simply a shifting alignment pattern in the field.

Imagine:

- Dropping multiple pebbles into a still lake.
- The ripples interact, interfere, and recombine.

To an unaware observer seeing these patterns at different locations and times, it may look like something is "moving." But in reality, nothing is moving. Only alignment patterns are reconfiguring.

So:

Motion = Interference of ripple alignments + Observer's frame of perception

No object is "carrying" motion.

Instead:

- It appears in one place when the ripple aligns.
- Then it appears in another — as the ripple alignment shifts.
- The field never moves — it re-expresses.

Time: The Consequence of Ripple Resolution Delay

Now let's bring in time.

In UED, ATT (Alignment Threshold Time) is the true underlying reason we perceive time.

What is ATT?

The delay between the ripple cause and its perceptual resolution.

It's not a flowing dimension. It's a consequence threshold — a ripple countdown.

Every event — from photon appearance to emotional memory — happens when the field aligns. Not before. Not after.

What you call "time passing" is simply:

- A series of resolved consequences
- Stacked into memory
- Sequenced through perception

Thus:

Time is not something flowing. Time is the perceptual spacing between ripple consequences.

Remove the observer → no time. Remove the ripple → no sequence. What remains? Stillness.

Motion and Time Are Field Responses, Not Cosmic Ingredients

Think of it like this:

- You're not "moving through time."
- The field is unfolding ripple alignments, and your perception is stringing them together.

You are not a body traveling. You are a participant in a ripple choreography that presents the illusion of motion.

What we call:

- Velocity → is alignment rate
- Acceleration → is field tension
- Distance → is alignment span

- Time → is ATT accumulation

These are not cosmic absolutes. They are field resolutions made visible.

Death: The Collapse of Time and Motion

Let's go deeper.

When a being dies:

- The ripples that once allowed perception collapse.
- The alignment between observer and environment dissolves.

What follows?

No time. No motion. No experience.

Because there is nothing left to participate in ripple resolution.

This is why UED considers:

- Time and motion are active consequences of conscious ripple participation.
- They don't exist for the dead. They don't pre-exist for the unborn.

They are activated phenomena — not eternal ingredients.

Why Time "Slows Down" or "Speeds Up"

Ever noticed how:

- Time feels slow in boredom?
- Time flies in excitement?
- Trauma feels frozen?
- Meditation suspends time altogether?

These aren't tricks of the mind. They are real variations in ripple participation.

When your perception aligns faster with ripple events, time compresses. When ripple alignment is delayed, time stretches.

In UED psychology: Attention is ripple alignment speed.

Thus:

- More aligned → denser time
- Less aligned → slower consequence

Time is not objective. It is a localized ripple rate.

Final Insight

So what are we saying? That the universe is not ticking. That space is not racing. That atoms aren't spinning in some cosmic stopwatch.

We're saying:

✗ Time is not flowing
✗ Motion is not occurring
✗ You are not "in" time or space
✓ You are experiencing ripple consequences
✓ You are resolving alignment across a still field
✓ You are perceiving, not traveling

So remember: What we call "time" is how long it takes for the field to say yes. What we call "motion" is where the field permits alignment next.

Everything else is inherited illusion.

Stillness is the truth. And in that stillness, everything happens — without ever moving.

20: Electromagnetic Field

The Ripple Grid Behind All Interactions

How UED Reveals the Deeper Pattern Beneath Force and Flow

We have reached the invisible fabric that underpins nearly everything modern physics calls *interaction* — the electromagnetic field.

In the classical framework, the electromagnetic field is described as:

- An entity extending through space,
- Governed by Maxwell's equations,
- Carrying energy and momentum through electric and magnetic vectors.

From atoms to antennas, from your heartbeat to gamma rays, it's all shaped by this field.

But mainstream science still imagines this field as an *active agent embedded in space* — oscillating, propagating, influencing charges — while space itself remains passive.

Universal Energy Dynamics (UED) breaks this illusion.

In UED, the electromagnetic field is not a separate entity "inside" space. It is the ripple grid of space — structured, stationary, and loaded with alignment potential.

It is not *in* space. It is the visible aspect of the structured vacuum energy field (SVE).

Let's decode.

Beyond Maxwell: The Static Substrate of Ripples

Maxwell's equations revolutionized our understanding of electromagnetism by linking electric and magnetic fields as interdependent waveforms, allowing predictions of light propagation and radiation.

But Maxwell began with an assumption:

That these waves travel through space — as if space were a medium like the 19th-century "aether".

After Einstein discarded the aether, physicists retained the equations — but silently replaced the medium with an empty space carrying mysterious fields.

UED questions both assumptions.

There is no need for a traveling wave. No need for fields moving through a vacuum. No need for energy being "sent" across distance.

Instead:

Every charge, current, or emission ripple-tunes the local field. Electromagnetic effects appear wherever alignment conditions are met, not where energy arrives.

This repositions the field:

- Not as a carrier, but as a canvas.
- Not as a force, but as a form of ripple coherence.

The Real Grid: Ripple Geometry and Pattern Locking

In UED, the electromagnetic field is understood as a ripple pattern formed by:

- Local field tension around particles or objects,
- Structural orientation (dipole, monopole, spin effects),
- Phase alignment with surrounding field nodes.

What we call:

- Electric field = directional ripple alignment with scalar potential gradients.
- Magnetic field = rotational ripple lock formed by charge participation in curved field tension.

No independent entities. No "magnetic lines" floating in air. Just ripple geometries, interpreted through instruments and models.

For example:

- A charged particle is not "surrounded" by a static electric field.
- It locks the field into a ripple bias — detectable through its consequences on other ripple participants.

Similarly:

- A moving charge doesn't "emit" a magnetic field.
- It introduces a spin-phase distortion in the structured field, which we interpret as magnetic influence.

Why Opposites Attract and Likes Repel — Ripple View

Conventional physics explains attraction and repulsion through field line compression and Coulomb's law.

But under UED:

- Like charges create field redundancy — ripple compression leads to repulsion as the field seeks tension release.
- Opposite charges create ripple complementarity — aligning to resolve field asymmetry, leading to attraction.

Thus:

Electrostatic "force" is just ripple field seeking harmonic equilibrium.

This reinterprets electric potential not as stored energy, but as stored ripple imbalance.

Why the Field is Still — Yet Everything Happens

If the field is still, how do electric and magnetic events occur?

Because "stillness" in UED doesn't mean nothingness — it means no intrinsic motion.

The field is fully structured — filled with potential ripple configurations.

When ripple events occur (like charge placement, current, or photonic disturbance), the field reconfigures locally.

This reconfiguration:

- Does not "move" outward.
- It resolves across pre-structured ripple links.

Imagine a stretched net:

- Pull one node → others adjust instantly within their tension constraints.
- No thread moves — but the pattern shifts.

Electromagnetic interactions are like this:

- Ripple triggers → field realigns → consequence manifests.

From Antenna to Atom: Everything Is a Ripple Participant

In UED view:

- An antenna resonates with ripple fields.
- A metal wire modulates the local ripple tension.
- A cell membrane interacts via micro dipole-induced ripple locks.
- A neuron fires by phase-shifting its local field enough to cross perceptual thresholds.

Thus: The electromagnetic field is not one field — it is a localized ripple state defined by the object's geometry, tension, and environmental alignment.

Photons Are Not Field Packets — They Are Consequence Pings

In quantum electrodynamics (QED), photons are virtual or real particles — "force carriers" between charges.

UED challenges this:

No particle is transferred. The "photon" is the name we give to a resolved ripple event between emitter and observer.

There's no photon flying through space. There's just the field resolving its ripple difference into an observable consequence.

This redefinition erases:

- Wave-particle duality,
- Photon path paradoxes,
- Non-local communication illusions.

Instead, everything becomes ripple alignment and consequence.

Electromagnetic Radiation Reimagined

Radiation is conventionally described as energy radiated from a source through electromagnetic waves.

UED reframes it: What we call "radiation" is the field's way of restoring ripple tension equilibrium — like a stone dropped into a pond creates field realignment ripples, not energy projectiles.

- The wave doesn't move — it appears wherever ripple threshold is met.
- No energy "travels." Instead, field zones resolve consequences at differing ATT (alignment threshold time).

This explains:

- Why electromagnetic waves can "pass through" transparent media,

- Why fields can "penetrate" walls (through ripple tolerance),
- Why coherence, not just intensity, determines perceptibility.

Final Insight: The Grid Is You

We often talk about the electromagnetic field as if it's "out there."

But every:

- Thought,
- Movement,
- Emotion,
- Perception...

...involves electromagnetic ripple alignment inside your own body and brain.

You are a ripple participant in the electromagnetic grid. You don't just observe the field — you modulate it.

UED doesn't just redefine physics. It redefines identity.

You are not separate from the field. You are its consequence — and its interface.

21: Field Healing

How Alignment Affects Health, Emotion, and Memory

Why your body isn't just a machine, but a ripple participant in a conscious field

Science has long pursued healing by treating the body like a machine:

- Diagnose the damaged part,
- Apply the right chemical,
- And wait for mechanical recovery.

But this machine model fails to explain why:

- Some people heal faster without medical intervention,
- Emotions trigger or resolve illness,
- Placebo works,
- And how memories, long buried, cause real physical symptoms.

These are not glitches. They are clues. Clues pointing toward something mainstream science refuses to see:

You are not a biochemical machine. You are a field participant. Healing is not just chemical — it is ripple re-alignment.

Let's unfold this new paradigm of healing — one ripple at a time.

The Body as a Ripple Structure, Not a Chemical Container

According to UED, your body is not just cells and molecules. It is:

- A condensed ripple arrangement,
- Interfacing with surrounding field structures,
- Constantly tuned by internal alignment and external resonance.

This ripple arrangement is dynamic. It's not fixed. It's always in flux, adjusting based on:

- Thoughts,
- Emotions,
- Environment,
- Belief,
- Sleep,
- Food,
- Breath,
- Relationships.

Health is not just the absence of symptoms. Health is stable, coherent ripple participation with the field.

When this participation gets misaligned, you experience:

- Fatigue,
- Anxiety,
- Inflammation,
- Mental fog,
- Disease.

Emotions: Ripple Turbulence in the Internal Field

Let's focus on emotions — often dismissed as "soft" or "subjective."

In UED, emotion is ripple turbulence caused by:

- Conflicting memories,
- Internalized perceptions,
- Anticipatory projections.

Each emotion is a pattern of ripple distortion in your local field:

- Grief: Compression without outlet.
- Anger: Unresolved field tension seeking external discharge.
- Anxiety: Anticipated misalignment with unknown consequences.
- Joy: Harmonized field resonance, typically non-local and expansive.

Emotions are not illusions. They are real field events, with real consequences.

These ripple events:

- Change body chemistry,
- Shift organ function,
- Create energetic feedback loops.

The body reacts — not because the mind imagines, but because the field actually shifts.

Memory: Frozen Ripples, Repeating in Loops

Memory is not stored in the brain like files in a hard drive. In UED:

Memory is a ripple lock — a standing field pattern held within the body's local ripple architecture.

Trauma, for instance, creates a high-tension ripple freeze that distorts:

- Perception,
- Response,
- Health.

This is why:

- A smell, word, or image can trigger a full-body reaction.
- People "relive" trauma — because the ripple is still active, waiting for resolution.

Healing requires not suppression, but ripple re-resolution.

That's why techniques like:

- Breathwork,
- Somatic therapy,
- Deep touch,
- Energy healing,
- Even storytelling and art...

...work. They create new ripple alignments that unlock old patterns and allow the field to re-harmonize.

Illness: Misalignment, Not Just Malfunction

In modern medicine:

- Illness is a mechanical breakdown or chemical imbalance.
- The solution is repair or replacement.

But UED asks:

What caused the imbalance? What misaligned the system?

From UED view:

- Illness is a field resonance breakdown.
- The organ or system falls out of tune with the larger ripple field (internal and external).
- The body tries to signal this misalignment through symptoms.

Example:

- Heart issues may emerge after prolonged grief — ripple compression in the chest region.
- Digestive disorders may arise from chronic control issues — misalignment in the solar plexus field.

- Skin problems often correlate with boundary perception distortions — ripple friction with outer field.

You don't just "catch" illness. You fall out of ripple coherence.

Healing = Re-Alignment

Real healing is not about attacking a virus or killing a cell.

It's about reintegrating your ripple structure with the universal field.

This involves:

- **Clarity**: Recognizing internal distortion patterns (mental, emotional, spiritual).
- **Stillness**: Creating ripple silence so new alignment can occur.
- **Flow**: Participating gently with nature, breath, water, movement.
- **Consequence awareness**: Understanding how unresolved patterns echo in the body.

In UED healing, nothing is forced.

Instead, the field is invited to:

- Re-settle,
- Re-align,
- Re-integrate.

The result? Sustainable healing — not suppression.

Placebo, Belief, and the Mind-Field Interface

Mainstream science struggles to explain why placebos often work — even when the subject knows it's fake.

UED resolves this clearly.

Belief = internal ripple reprogramming. The body-field responds not to the pill — but to field coherence restored by expectation alignment.

When you believe healing will occur:

- Ripple readiness improves.
- Tension softens.
- Alignment threshold is met.
- Healing resolves.

So placebo is not trickery — it's a demonstration of how powerful internal field participation is.

This is why:

- Faith heals.
- Hope relieves.
- Forgiveness cures.

These are not moral virtues — they are ripple tools.

Final Insight: You Are the Field You Heal In

We've looked at the body as a ripple matrix. We've seen emotions as turbulence, memory as field lock, and illness as misalignment.

So what is healing?

Healing is the re-harmonization of the self with the still field of existence. It is not something done to you — it is something allowed through you.

You're not just a patient. You're a ripple participant with agency to:

- Listen,
- Align,
- Resolve.

You don't have to wait for medicine. You can tune your field — through breath, rest, silence, love, forgiveness, and presence.

You are not a machine to be fixed. You are a ripple alignment seeking stillness.

And the moment stillness returns — healing begins.

22: Consciousness

The Field That Knows Itself

You are not inside the universe — you are its ripple realization

If the Earth is still, And if space doesn't expand, And if light doesn't travel, And if motion is just ripple participation...

Then what, exactly, is *moving* through experience?

The answer, surprisingly, is not "you" — as a body, identity, or thinker. The answer is: consciousness — the field noticing itself through ripple consequence.

We've questioned the mechanics. We've challenged the frames. We've redefined light, mass, and energy.

Now it's time to ask:

Who is asking all this?

Consciousness Is Not In You — You Are In Consciousness

The modern paradigm sees consciousness as a product of the brain:

- Neurons fire,
- Signals emerge,
- Somehow, *experience* arises.

This reductionist view leads to a dead end:

- No one can locate where memory "resides."
- No scan shows "thinking" happening as a location.
- No biology has explained the sense of I.

UED offers a radically different model.

Consciousness is not *generated*. It is fundamental. It is not *personal*. It is participatory.

You do not "have" consciousness. You are a ripple interface through which universal consciousness aligns and expresses.

Just as a ripple appears in a pond — not separate from the water — you appear as experience — not separate from the field.

The Still Field is Conscious — That's Why It Can Be Known

UED proposes that the structured vacuum energy (SVE) isn't just a substrate — It is intelligent equilibrium.

- Not thinking.
- Not reacting.
- But aware by design.

This awareness is not linear or limited. It is:

- Non-local,
- Present everywhere simultaneously,
- Capable of forming ripples that observe each other.

What we call:

- Self-awareness,
- Intuition,
- Insight,
- Knowing without learning...

...are all signs that we don't need to install consciousness into matter — We simply need to realize that matter is a ripple within it.

Perception = Alignment Threshold Consequence

In previous chapters, we defined ATT (Alignment Threshold Time) as the key to when a ripple interaction becomes visible or perceptible.

Here's the deeper truth:

ATT is not just a physical phenomenon — it is a conscious selection of consequence.

Perception is not passive:

- Your consciousness "locks" into certain field conditions,
- Resolves a ripple,
- And creates the experience of reality.

This explains why:

- Two people experience the same event differently.
- Dreams feel real.
- Thoughts affect physiology.
- Placebo works.
- Near-death experiences show clarity while the body is "dead."

Because perception is not data collection — it is conscious field interaction.

You are the ripple resolving *yourself.*

Mind Is the Local Interface; Consciousness Is the Field Itself

Let's make a sharp distinction.

- Mind = ripple consequence manager (temporary, patterned, memory-based).
- Consciousness = field awareness (non-local, timeless, structure-aware).

The mind thinks it is "you," But the moment silence happens — consciousness takes over.

This is why:

- In meditation, thinking stops, but awareness remains.
- Under anesthesia, the mind disappears, but some part remains present.
- In flow states, identity fades, but intelligence acts precisely.

Because mind is a tool, But consciousness is the canvas — and the painter.

Consciousness Knows the Still Earth

The ultimate irony?

Mainstream science can model galaxies, spin black holes, and accelerate particles, But it still can't explain why you feel present here.

UED turns the question inside out:

You feel still — because you are aligned with the still field. Your perception of Earth's stillness is not delusion — it is direct awareness of equilibrium.

Consciousness doesn't move — it perceives ripple movement. Consciousness doesn't orbit — it observes alignment from a center.

So, when you say, "The Earth feels still to me," You are not being naive — you are being honest.

Your consciousness is reporting alignment, not measuring motion.

Final Insight: You Are the Field, Not the Fleeting Form

You are not a by-product of a lucky DNA chain. You are not a temporary trick of evolution.

You are ripple coherence in a knowing field. You are the interface of alignment and intention. You are consciousness experiencing consequence.

The Earth doesn't move — because *you*, at your truest level, don't move either. Only ripples do.

And when you remember this —You heal. You clarify. You live in alignment.

Because when the field knows itself through you, Stillness is no longer a theory — it becomes a state of being.

Prelude: When We Finally Looked Up

For centuries, we've been told the universe is rushing. Expanding, spinning, stretching, accelerating, collapsing, bouncing back. But all this rush was never seen. It was interpreted.

We see the stars. We don't see their motion.

Night after night, generation after generation, the stars appear fixed in the sky — anchored like cosmic sentinels in a shared ripple field.

Even galaxies beyond naked-eye range reflect the same truth when instruments are stripped of assumed motion logic:

- No one ever saw expansion — they saw redshift and imagined it as motion.
- No one ever saw rotation — they saw patterns and turned them into spins.
- We never saw Earth move — we inherited the idea.

We trusted indirect effects — like day-night cycle, seasons, or Doppler distortions — as if they were evidence of motion.

But these were ripples of interaction, not proof of displacement.

We didn't observe motion — we wrote it into the sky.

The Earth may feel still because it *is*. The stars may appear still because they *are*. Your perception isn't failing — it's reporting alignment, not illusion.

So, in this final part, we'll dare to do what few ever tried:

Strip away every projection, Return to stillness, And look again — not with eyes clouded by equations, but with clarity born of consciousness.

Welcome to the end of motion. And the beginning of perception.

Part IV – Cosmology, Perception, and Reality

23: The Big Bang Isn't a Beginning

Field Tension, Ripple Release, and Cosmic Phase Change

In the beginning, is not a scientific claim. It is a human need for story.

For centuries, the Big Bang has been treated as the origin of time, space, matter, and all known phenomena. According to mainstream physics, all energy and mass were once compressed into a singularity—an infinitely dense, infinitely hot point that exploded approximately 13.8 billion years ago, giving birth to the universe.

But what if this idea was never a true beginning, but simply a ripple consequence misread as creation?

What if the universe didn't begin with a bang at all—but with field tension resolving through structured release?

Field Equilibrium Before "Time"

In the Universal Energy Dynamics (UED) view, the true default state of the cosmos is stillness —a structured vacuum field in equilibrium, free from directional flow or temporal sequence. This isn't emptiness; it's a fully potentialized yet undisturbed field.

Time, motion, matter, and energy are not foundational aspects of the universe. They are resolved ripple events that emerge after equilibrium is disturbed.

So, the question isn't "What caused the universe to begin?" but rather:

What caused the field to shift out of perfect balance?

This shift, in UED, isn't a bang. It's a cosmic re-tuning—a phase change from pure stillness to ripple tension.

Cosmic Ripple Pressure: The Real Genesis

Just as ripples form on a still pond when a stone disturbs its surface, the structured vacuum field reacts to imbalances through ripple propagation. But unlike water, the field doesn't need a foreign object to begin rippling.

Ripple tension can build up internally, due to asymmetry or imbalance in the field configuration. When the potential becomes unsustainable, the field doesn't explode—it resolves.

This resolution appears to an observer (like us) as an expansion, or what science interpreted as the Big Bang.

But what if that expansion was not outward movement, but multi-directional ripple unfolding?

Just like energy discharges from a compressed spring, the vacuum field may have released tension via structured ripple flows — giving birth to what we later interpreted as light, mass, and motion.

Phase Change, Not Creation

The Big Bang, then, is not a creation event. It is a ripple consequence of a phase transition in the cosmic field:

- From undisturbed stillness to ripple activation
- From potential energy to ripple dynamics
- From no-time to ATT (Alignment Threshold Time) boundaries

This model doesn't deny early-universe observations such as background radiation or redshift. It simply re-interprets them:

- **Cosmic Microwave Background**: Not a leftover from a fiery explosion, but a uniform ripple residue of large-scale field alignment.
- **Redshift**: Not galaxies flying away, but ripple delay effects due to field re-stabilization.
- **Large-scale structure**: Not chaotic gravitational clumps, but emergent patterns from ripple synchronization.

No Singularity. No Explosion. Just Ripple Logic.

The idea of an infinitely dense singularity has always been a mathematical crutch, not a physical necessity. UED removes the need for this abstraction entirely.

There was no point-like origin. There was field resolution:

- Ripple consequence spreading simultaneously across directions.
- Field equilibrium seeking balance through phase equalization.
- A cosmic unfolding, not a cosmic birth.

Time didn't start. Motion didn't begin.

Perception began—once ripple consequence crossed the threshold of alignment.

And the rest is not history. It's field response.

Conclusion: The Universe Never Began — It Transitioned

The Big Bang isn't the origin of the universe. It's the observable start of imbalance resolution.

- No fixed point.
- No singular cause.
- No ticking clock.

Just a still universe reconfiguring itself through structured ripples.

And what we call "everything" is simply the echo.

Next, we turn to the most celebrated distortion of all: Relativity. What Einstein saw was brilliant, but what he missed changes everything.

24: What Einstein Missed

Relativity Relativized

Time dilation as ripple threshold, not space-time warp

Einstein revolutionized physics with his theory of relativity by claiming that time and space are not absolute, but interwoven, and influenced by mass and velocity. The faster you move, the slower time flows. The more massive an object, the more it bends space-time. This model fundamentally changed our understanding of the cosmos.

But despite its brilliance, relativity carries within it a hidden assumption that has gone unchallenged:

That time is an entity that can be warped or stretched.

In UED, this interpretation is not only unnecessary—it is misleading.

Time Dilation Reframed

According to Einstein, an astronaut traveling at near-light speed would experience time more slowly than someone on Earth. This phenomenon, confirmed by experiments with atomic clocks on fast-moving jets and satellites, is called time dilation.

UED does not deny this measurable difference. But it reframes it:

Time dilation is not due to a "warped space-time," but a delay in ripple alignment crossing ATT (Alignment Threshold Time).

Motion, in UED, is not real displacement through space. It is a reconfiguration of ripple fields. As relative ripple participation changes, so does the timing of perceptual resolution. That delay is what we interpret as "slowed time."

In this view:

- Time is not stretched.
- Clocks don't tick slower because of velocity.
- The ticking itself is a ripple event, and the delay is a field effect.

Clocks Measure Ripple Events

All timekeeping devices are ultimately built on periodic phenomena: vibrations, oscillations, decay rates. But each of these depends on ripple interactions being resolved at consistent ATT thresholds.

If the ripple field is stretched, compressed, or reconfigured due to relative motion or gravity, ATT thresholds shift.

So, what we measure as time dilation is simply this:

- A ripple interface is reaching threshold resolution slower due to altered participation.
- Time appears stretched, but in reality, the field is taking longer to resolve the same ripple consequence.

This dissolves the need for mysterious fabric-warping or 4D geometries.

Relativity vs Reality

Einstein saw a universe where energy and matter mold the space-time fabric, and observers experience reality differently based on speed and mass. But UED sees a universe where:

- There is no fabric.
- There is no warping.
- There are only ripple thresholds.

Everything else is an observer-based delay in resolution, not a real change in the universe.

This restores simplicity:

No need to twist reality—just follow the ripples.

Einstein's Genius, and His Gap

Einstein was right to question Newton. He was right to emphasize the role of the observer. But he made one key mistake:

He materialized perception, turning the experience of delay into a structure of warping.

What he called space-time was simply a mapping of ripple consequences. And what he called time dilation was a delay in field resonance.

Einstein looked into the field, but stopped short of asking:

What causes this ripple consequence to appear when it does?

UED answers that: Alignment. And with alignment, time reappears as what it has always been:

- Not a flowing dimension,
- But a threshold signal in a still universe.

25: Quantum Myths

Why Observation Changes Reality

Ripple collapse and ATT zones

The Quantum Confusion

From the double-slit experiment to Schrödinger's cat, quantum physics has mesmerized generations with a single paradox:

How can mere observation change reality?

Electrons behave like waves when unobserved, yet act like particles when measured. Photons "decide" whether to interfere or not based on whether you're watching. Entanglement connects particles across space instantly. These are not just experimental anomalies; they have been elevated into *mystical truths*—that consciousness affects matter, that probability is more real than reality, and that everything is uncertain unless observed.

But this magical thinking is the result of a misunderstanding. Quantum physics is not magical—it's incomplete.

UED brings clarity: the confusion arises not from the experiment, but from how we interpret the act of observation itself.

The Observer Is Not a Wizard

In traditional interpretations, an observer seems to possess strange powers. Before you look, the system is undefined—existing in a "superposition." After you look, it "collapses" into a definite state.

This is backward.

The observer is not causing reality. The observer is waiting for ripple resolution. And ripple resolution is a field alignment process, not a mystical act.

In UED, what's called "collapse of the wave function" is just this:

Ripple alignment reaching ATT (Alignment Threshold Time).

Observation doesn't *change* the electron—it reveals it.

Before alignment, ripple interactions are real, but not visible. After alignment, they resolve as observable consequences—like a spot on the screen, a click in the detector, or a measurement on the device.

Superposition = Pre-Resolution State

In UED, "superposition" is not a particle being in two places at once. It is the field holding multiple ripple possibilities, none of which have crossed the ATT threshold yet.

It's not uncertainty—it's unresolved consequence.

The moment alignment occurs with the observer's field (instrument, brain, etc.), the consequence appears.

Thus:

- A photon does not choose a path.
- An electron does not split or merge.

- Schrödinger's cat is not dead-and-alive.

These are interpretive artifacts of viewing ripples as particles, and ATT zones as sudden changes.

There is no "collapse"—only field coherence reaching perceptual threshold.

Why Observation Matters—But Isn't Magic

Observation *seems* to change reality because the observer is a participant in the field. Their own ripple signature interacts with the system. The detection device, the experimenter's presence, and the alignment setup all contribute to whether and when ATT is crossed.

That's why:

- No observation = no resolution.
- Observation = threshold crossing = resolved ripple.
- Different setups = different resolved consequences.

This isn't magic. It's consequence logic.

The Quantum Measurement Problem—Solved

In mainstream science, measurement causes collapse. In UED, measurement is ripple convergence.

The so-called "observer effect" is not about human consciousness warping probability. It is about field dynamics converging into a visible, measurable alignment.

Every observation is simply the universe whispering:

"The ripple has now reached you."

And everything prior to that is not unknowable—it is unresolved.

Conclusion: You Never Created Reality—You Aligned With It

Quantum mechanics baffled scientists because they couldn't accept that observation doesn't *create* events—it *times* them.

The real question is not *why does the electron collapse into one position?* But rather:

Why did your shared ripple field resolve that outcome at that moment?

UED answers: Because that was the moment ATT was reached. Because that was when ripple alignment allowed perception.

Not magic. Not mystery.

Just field logic.

26: The Observer Illusion

You Are the Ripple

Conscious alignment with structured field

Who Is Observing?

Quantum physics made a profound admission: without the observer, there is no reality as we know it. Yet it never asked the deeper question—

Who exactly is this observer?

Most scientific models separate the observer from the observed—as if the human mind floats above nature, peeking into a universe that exists "out there."

But UED boldly dissolves this illusion:

The observer is not a separate entity. The observer is a ripple consequence within the field—just like everything else.

Your mind, your awareness, your conscious experience—they are not floating ghosts in a deterministic machine. They are structured ripple patterns inside the same field you are observing.

You are not watching the universe from outside. You are the field observing itself.

Consciousness = Ripple Alignment

Consciousness is not a magical spark. It is the ability of the ripple pattern to maintain coherent self-alignment within the structured vacuum field.

When ripple alignment is coherent enough, it forms a stable identity we call "I."

That "I" is not made of atoms, DNA, or brain cells. It is the observer ripple—a self-resonant pattern that can tune into other ripple configurations in the field.

So when you "observe" a photon, an apple, a star, or an idea—what's really happening is:

- Your ripple field is intersecting with another
- When ATT (Alignment Threshold Time) is reached, consequence is resolved
- And that resolution is what we call perception

You don't observe the world. You resonate with its consequences.

There Is No Outside

The illusion of "me here" and "world out there" arises from spatial consequence—the arrangement of resolved ripples. But in the field itself, there is no boundary.

- Your body is a ripple knot
- Your brain is a processing echo
- Your thoughts are localized field modulations
- Your memories are ripple trails

You don't exist *within* the field. You are the field—folded into a ripple form that can reference itself.

And the act of observation is just one ripple meeting another.

Stillness: The Ultimate Witness

All ripples eventually seek rest. That is the nature of the field. Motion is temporary. Turbulence resolves. Alignment always tries to restore stillness.

Consciousness—true, unshaken awareness—is that point of still alignment where:

- Observer and observed collapse into unity
- Thought gives way to presence
- Reaction becomes recognition

That moment is not a peak of mental activity—it is the zero-point of ripple dissonance.

You, the observer, are not a puppeteer pulling the strings of nature. You are the string vibrating in harmony—or dissonance—with it. And when fully aligned, you are not even observing. You are simply being.

The Final Insight

The famous quantum quote says: *"The observer creates reality."* But in UED, we revise it: The observer is the ripple by which reality aligns itself.

You didn't come from outside the universe to look at it. You were always a ripple in its stillness—remembering itself.

You are the wave that learned to watch. The field that learned to feel. The stillness that remembers motion.

Not a spectator. A participant. Not the mind inside the body. But the ripple inside the whole.

27: Stillness Is the Final Law

All motion seeks ripple resolution to stillness

Motion: The Greatest Misunderstanding

Physics begins with motion. From Newton's laws to Einstein's spacetime curvature, from quarks zipping in quantum fields to galaxies hurtling across the cosmos—motion is the language of modern science.

But this obsession with movement hides a deeper truth.

Motion is not the origin. Motion is the symptom of imbalance.

And like all symptoms, motion doesn't last forever. It is temporary. It seeks resolution.

At the heart of the universe lies not chaos, but order. Not movement, but stillness.

And that stillness is not the absence of action—it is the completion of it.

Ripple Resolution: The Path Back to Equilibrium

Under Universal Energy Dynamics (UED), everything that moves is a ripple consequence. A ripple is not "something" moving through "nothing." It is a local imbalance in the field—a structured deviation from stillness.

And what does every ripple do?

It spreads... it interacts... it echoes... And finally—it dissolves.

Like a drop in a pond, the ripple doesn't create motion endlessly. It seeks the lowest tension, the smoothest curve, the gentlest path back to equilibrium.

This is not metaphor—it is field mechanics.

Whether it's a falling apple, an orbiting moon, a burning star, or a single photon:

Every motion is a field attempting to return to its still state.

Stillness Is Not Inaction—It Is Perfection

Stillness does not mean "nothing is happening." It means:

- All ripple interactions are balanced
- No direction dominates
- No unresolved energy remains
- No perception threshold is crossed

In UED, this is the natural resting state of the universe.

Motion is what happens when equilibrium is broken. And that motion continues until stillness is restored.

In other words, all physical processes—whether thermodynamic, mechanical, gravitational, or electromagnetic—are just various forms of ripple resolution.

Stillness is the final law, because it is the default state every system tries to return to.

Even Time Seeks Stillness

Time, as defined in UED, is not a flowing dimension. It is a threshold artifact—a perceptual record of ripple transitions.

The more misaligned the system, the more intense the ripple activity—hence, the more *events* and the stronger the illusion of time.

But as field alignment increases, perceived time fades:

- In deep meditation: time slows down
- In death: time stops altogether
- In moments of clarity: time feels suspended
- In the structured vacuum: there is no time at all

This is not poetry. It is the natural de-escalation of ripple tension.

Time itself wants to resolve into stillness.

The Universe Is Not Expanding Forever

Many scientists believe the universe will either expand endlessly, collapse back in on itself, or freeze into entropy.

But from a UED perspective, expansion is not direction—it is dissipation.

The cosmos is not rushing outward. It is rippling toward rest.

All energy disperses. All ripple memory fades. All motion dilutes... until what remains is quiet.

Not annihilation. Not heat death.

But balance.

Stillness.

You Are the Echo of Stillness Trying to Return

Your thoughts, your movements, your desires—they're ripples too.

- When your mind is restless, it is in imbalance.

- When your emotions flare, your field is disturbed.
- When your body is sick, it seeks realignment.

And what brings healing? Clarity? Peace?

Stillness.

At every scale—from quantum to cosmic, from biological to emotional—the field seeks one thing:

- Resolution into equilibrium.
- Restoration of coherence.
- The return to stillness.

The Final Law

Not force. Not mass. Not velocity. Not relativity. Not entropy.

The final law is this:

All motion exists to resolve itself into stillness. Stillness is not what follows the end. It is what was always there before the beginning.

That is not a philosophical idea. It is the mechanical truth of the field.

Stillness is not the destination of the universe. Stillness is the universe—patiently waiting for all things to return.

28: The Return to Common Sense – But Sharper

Giving power back to perception with clarity

Common Sense Was Never the Enemy

Modern science prides itself on going beyond "common sense." It looks down upon intuition, mocks the simplicity of direct perception, and claims that truth is often counterintuitive.

We were told:

- The Earth spins, even if it doesn't feel like it.
- Time bends, even if clocks don't show it.
- Objects exist in multiple places until we look.
- Nothing is at rest—everything is in motion.
- You must trust the math, not your eyes.

And so we obeyed.

We surrendered perception. We outsourced understanding. We were taught to doubt our senses, and revere abstract models instead.

But what if perception was never wrong? What if common sense didn't fail—what if it was simply interrupted?

Perception Is a Field Instrument

Your body is not a passive lump of meat. Your senses are not illusions. Your consciousness is not an accident.

You are a ripple participant in the field. And your perception is your primary detector.

It's not primitive—it's aligned. It doesn't deceive—it reports what's resolved.

The reason you feel Earth is still...the stars are fixed...the sun rises and sets...is because those are the field alignments resolved through you. Not hallucinations. Not delusions. But the actual resolved reality at your position in the ripple field.

Common sense is not obsolete.

It is simply the field's native language spoken through your alignment.

Clarity Is Not Simplicity—It's Precision

But don't confuse "common sense" with being simplistic. We don't return to flat-Earth myths or naive superstition. We're not anti-science. We're pre-bias.

This return is not to primitive thinking. It is to pre-assumption clarity.

The sharpened mind does not ignore its senses—it refines their meaning.

When you perceive stillness, you no longer ask: *"Why don't I feel motion?"* You ask: *"Is motion really happening, or just assumed?"*

When you see day turn to night, you don't think: *"Earth must be spinning!"* You consider: *"Could this be a field rhythm interpreted as rotation?"*

When you observe anything, you pause and ask:

"Is this a consequence of ripple resolution, or a model I've inherited?"

That's not ignorance. That's intelligent perception.

The New Scientific Method: Perceive, Align, Then Measure

We don't reject instruments or equations. We reject blindness.

UED proposes a new order:

1. Perceive without bias
2. Check alignment across field zones
3. Only then apply math and model

Because perception is not noise—it is the first ripple contact.

You are the detector. You are the observer. You are the first layer of physics.

From Wonder to Wisdom

When ancient humans looked at the stars, they saw stillness. Not because they were stupid—but because they were right.

They sensed rhythm, not rotation. Cycles, not spin. Order, not chaos.

And over time, we traded wonder for theory. We replaced direct alignment with indirect interpretation.

But now, the circle is closing.

We return to perception—but with knowledge. We return to simplicity—but with depth. We return to stillness—but with awareness.

This is not the end of science. It is the beginning of intelligent observation.

Final Reminder

You are not asked to believe a new model. You are invited to see again—with clarity.

Let the math sit quietly. Let the dogmas fall aside. Let your mind be still...and just watch:

- The stars that haven't moved.
- The Earth that doesn't tremble.
- The sky that breathes in rhythm.
- And the stillness that was always there, behind the noise.

Welcome back. To common sense. But sharper.

Epilogue: The Stillness We Forgot

In a world obsessed with velocity, orbits, and equations, stillness has become taboo.

Motion became truth. Stillness became denial. But beneath the grand narratives of spinning planets and expanding universes, something timeless awaited recognition:

The Earth never moved. We did—in thought, in theory, and in assumption.

Every law that demanded acceleration, every frame that required relativity, every force that needed mass—was built atop a scaffold that never questioned its ground.

But what if that ground—the still field—was the only real thing all along?

This book was never a rebellion against science. It was an invitation back to observation, alignment, and common sense sharpened by consciousness.

We questioned Newton not to diminish him—but to rescue his insights from misapplied permanence. We reinterpreted Einstein not to erase his brilliance—but to reveal what he could not yet perceive.

You don't need to accept this framework today. But you cannot walk away untouched by it.

Because once you see the stars as still—not moving, not expanding, just there— Once you recognize the rhythm of day and night without spinning— Once you feel your body, your breath, and your awareness resting inside a ripple field—

Then stillness returns.

Not as philosophy. Not as mysticism. But as the final law of consequence.

In the beginning, we moved around the Earth. Then the Earth moved around the Sun. Now it is time to return to the truth:

Nothing moves. Everything aligns.

And in that alignment, We find peace. We find perception. We find the Universe.

Still. Silent. Perfect.

Author Reflection

This book was not born from rebellion—it emerged from stillness.

For years, like many others, I taught and accepted the rotating Earth, the expanding universe, and the glorious mathematics that made sense of it all. But something always whispered from within: a contradiction between what we see... and what we are told.

When I paused to truly observe—not just as a physicist or a thinker, but as a participant in existence—I found something radical:

The universe is not rushing. It is not spinning. It is simply **resolving**.

This reflection is not a denial of science—it is an extension of it. A bold continuation. A demand that science returns to its first instrument: the **observer**, and its first evidence: **perception**.

I did not write this to claim final answers. I wrote this to invite better questions.

If a young student, a seasoned teacher, or a quiet seeker finds even a flicker of awakening through these pages, the effort was worth it. If a scientist reconsiders the unquestioned assumptions beneath their equations, I have done my job. If even one reader stops and watches the sky—not with inherited belief, but with open presence—then this book has fulfilled its ripple.

To those who trolled, debated, or dismissed—thank you. You were part of the consequence chain.

To those who resonated—stay aligned. The stillness was never outside. It was always waiting within.

— **Sandeep Chavan**

Independent Researcher, Observer, and Ripple Among Ripples

Appendices

Appendix A: Glossary of Classical vs UED Terms

A side-by-side comparison for readers transitioning from conventional physics to the ripple-based framework of Universal Energy Dynamics (UED).

Classical Term	Definition in Conventional Science	UED Reinterpretation
Inertia	The resistance of an object to a change in its motion or state.	Ripple-lock alignment with a still field; not resistance but ripple equilibrium in local field tension.
Force	A push or pull acting on an object, causing it to accelerate (F = ma).	A symptom, not a cause—ripple imbalance trying to resolve field misalignment.
Mass	A measure of an object's resistance to acceleration; source of gravitational attraction.	Condensed ripple configuration; not intrinsic but a consequence of energy structured in the field.
Energy	The capacity to do work; exists in various forms (kinetic, potential, etc.).	Ripple displacement intent; not work-potential, but alignment tension within the field.

Classical Term	Definition in Conventional Science	UED Reinterpretation
Momentum	Quantity of motion, defined as mass × velocity.	Stored ripple potential; a memory of ripple displacement in the field.
Time	A continuous, one-directional dimension measured by clocks.	A threshold of ripple perception (ATT); emerges from the resolution rate of ripple consequences.
Space	A static 3D container where events unfold.	Not a container but a fielded stillness—structured zones of ripple readiness.
Frame of Reference	A coordinate system to observe motion.	A perceptual ripple position; all frames assume stillness but fail to admit it.
Gravity	An attractive force between masses, modeled as curvature of space-time.	Ripple pressure and field tension; mass alignment disrupts equilibrium, generating localized compression.
Acceleration	Rate of change of velocity; caused by unbalanced forces.	A retuning of ripple flow; not just vector change but directional reorganization of ripple alignment.
Light	A particle-wave duality entity that travels at fixed speed in vacuum.	A ripple appearance; it doesn't travel, but aligns to ATT when perception is resolved.

Classical Term	Definition in Conventional Science	UED Reinterpretation
Quantum Collapse	The process by which a quantum state becomes definite upon observation.	Ripple threshold locking into a consequence zone; not collapse but field resolution.
Observer	An external entity measuring a system.	A ripple within the same field, not external—observer is part of the observed.
Big Bang	The explosive origin of the universe from a singularity.	A ripple phase transition—not a beginning but a release of field tension into structured dynamics.
Relativity	Space and time bend under motion and mass; no absolute frame exists.	Misinterpretation of ripple delay as time dilation; stillness is the only absolute.
Wave-Particle Duality	A paradox of quantum physics where particles behave like waves.	A misunderstanding—only ripples exist, consequence appears as discrete when alignment is resolved.
Vacuum	Empty space, devoid of matter.	Not empty—structured with dormant ripple potential ready to align.
Consciousness	An emergent property of complex brains.	A ripple-aware self-alignment; the field remembering.

Final Note

This glossary is not a rejection of the classical terms—it is a reframing of their deeper meanings. UED does not destroy physics; it liberates it from assumptions and restores its pulse: observation aligned with reality.

If you find tension between both sides of this table—good. That's the ripple starting to move.

Appendix B: Newton's Own Doubts and Letters

A glimpse into the mind of the man who shaped classical mechanics—and the quiet tremors of doubt he left behind.

Sir Isaac Newton's Principia Mathematica gave the world a mathematical scaffold to interpret motion, gravity, and planetary dynamics. Yet, hidden beneath the precision of his laws was a humble scientist wrestling with the very assumptions he laid down.

This appendix draws attention to Newton's personal letters, marginal notes, and speculative thoughts—revealing that even he was not fully convinced about the universality of his own constructs.

1. Newton's Worry about "Action at a Distance"

In his famous correspondence with Richard Bentley in 1692–93, Newton openly expressed concern about gravity's mechanism:

"That gravity should be innate, inherent and essential to matter, so that one body may act upon another at a distance through a vacuum, without the mediation of anything else... is to me so great an absurdity that I believe no man who has in philosophical matters a competent faculty of thinking can ever fall into it."

Here, Newton himself admitted that he did not believe gravity worked "magically" across empty space. He suspected there must be some medium or mechanism—but could not define it.

UED takes this whisper seriously. The "vacuum" is not empty—it is a structured ripple field, and gravity is not a pull but pressure from ripple tension.

2. His Silent Struggle with Absolute vs Relative Space

Newton introduced the idea of absolute space—a metaphysical stillness against which all motion could be measured. But he also knew this was untestable. In private letters and drafts, he admitted:

"We are to admit no more causes of natural things than such as are both true and sufficient to explain their appearances."

This is a paradox. Newton used absolute space as a mental backdrop but knew it could not be observed.

Ironically, UED reinstates this absolute stillness—not as abstract, but as the physical field of structured energy, the only true frame where consequences are born and resolved.

3. Newton's Interest in Aether

Though not widely emphasized in modern physics education, Newton explored aether theories in his *Opticks* (1704), writing:

"Doth not this aethereal medium in passing out of water, glass, crystal, and other compact and dense bodies in empty spaces, grow denser and denser by degrees?"

He theorized that light and gravity might require a subtle medium for propagation.

UED boldly reclaims this hint: the "aether" Newton imagined was not wrong, it was just unresolved. The Structured Vacuum Energy (SVE) of UED is that missing mediator—not a fluid, but a ripple lattice of dormant intelligence.

4. Newton's Final Reflections

In the preface to the second edition of *Principia*, Newton humbly stated:

"I have not as yet been able to deduce from phenomena the reason for these properties of gravity, and I do not feign hypotheses."

He famously refused to "frame hypotheses" without sufficient cause—a caution echoed throughout UED's rigor.

Newton did not claim certainty. He admitted gaps.

Modern science often forgets this. It teaches Newton's laws as final, not foundational questions dressed as conclusions.

Closing Reflection

Newton opened the door. He was a pioneer, not a prophet.

This appendix is not meant to diminish his genius—but to humanize it. To show that even he doubted the very simplicity that later scientists worshipped as gospel.

The man who gave us gravity was unsure what gravity truly was. The man who built mechanics feared the vacuum might be more than empty.

In these honest admissions, we find the courage to move forward—not by rejecting Newton, but by completing his thought with a field theory he never had the tools to build.

We don't contradict Newton. We ripple past him.

Appendix C: Suggested Experiments

& Ripple Thought Tests

From tabletop reflections to cosmic perception—how you can engage with the stillness and ripple field yourself.

This book is not just an argument—it is an invitation. An invitation to experiment, to perceive differently, to *observe stillness* not with massive instruments or billion-dollar colliders, but with the instrument of attention.

You don't need a lab. You need clarity, patience, and a willingness to reawaken the original scientist in you—the child who looked up and asked, *"Why doesn't the sky move?"*

A. Ripple-Based Field Awareness Exercises (At Home)

1. The Still Cup Test

Setup: Fill a transparent glass with water. Let it rest undisturbed.

Observation: You will notice that after some time, even micro-ripples vanish. The surface becomes still.

Insight (UED): Stillness is not lifeless—it is the natural resolution of motion. Everything seeks ripple equilibrium. This parallels how all celestial alignments stabilize in still vacuum fields.

2. Shadow Tracking (Noon's Deception)

Setup: Place a vertical stick in the ground. Track the shadow from morning to evening.

Classical Interpretation: The shadow "moves" due to Earth's spin.

Ripple Reframe: You remain still in a ripple field; light angle changes due to field alignment dynamics—not global spin. The shadow is a local ripple consequence, not proof of Earth's rotation.

3. Gyroscope Revisited

Setup: Use a handheld gyroscope or spinning bicycle wheel.

Observation: The gyroscope maintains its orientation due to angular momentum.

Reinterpretation (UED): Rather than inertia from rotation, the gyroscope locks to local field orientation—its resistance is due to ripple-lock, not spin.

You can test this in moving vehicles, elevators, or while changing direction.

B. Field Tests and Observational Inversions (Philosophical + Physical)

4. The Reverse Orbit Thought Test

If Earth truly revolved around the Sun, stars would not appear fixed over centuries. Constellations should *warp subtly* over millennia.

Yet, ancient star maps *match today's sky.*

Test the Claim: Compare old sky charts (available in public records or planetarium archives) with present ones.

Ripple Interpretation: The stars aren't moving. We're not orbiting. The entire narrative of revolution may be field misinterpretation.

5. The Spin Test: Weather & Motion

If Earth spun at 1600 km/h at the equator, you should feel inertial effects—especially in eastward vs westward flight durations, or projectile trajectories.

But:

- East-west flight time differences are due to wind patterns, not Coriolis effects.
- Long-range artillery is corrected manually—not automated by rotation math.
- Sniper manuals mention Earth spin inconsistently.

Ask: Why are "corrections" not natural?

Ripple View: There's no spin—just ripple alignment with large field domains (atmosphere included).

6. ATT Awareness Test (Alignment Threshold Time)

Setup: Enter a dark room. Light a small candle. Observe when the light becomes "visible."

What changes? Not the light, but your field threshold. Your eyes, brain, and environment synchronize—then you "see."

Ripple Insight: Vision doesn't catch "traveling photons." It detects ripple resolution at perceptual threshold (ATT).

C. Observer Alignment Exercises (Mental Field Tuning)

7. Breath and Stillness Sync

Close your eyes. Inhale deeply, and slowly exhale. Stop trying to think. Just observe breath rippling through stillness.

Ask: *Where does breath go when it's not in motion?* Then ask: *Where do thoughts arise from, and where do they return?*

Insight: Even internal waves obey UED. Stillness is the field source—everything emerges and returns to it.

8. Consciousness Alignment Loop

Think of a past event. Now remember how your body felt.

Realize: That memory is stored ripple potential. You are not thinking—you are accessing ripple impressions that re-align your field.

Repeat this with sound, smell, or visual cues.

Conclusion: Memory is not internal data. It is field alignment recall.

D. Macro-Scale Observations to Reconsider

- Stars Don't Drift – Even over centuries, stellar positions remain coherent.
- The Moon Faces Us – It's always aligned, never rotating with respect to us.
- Seasonal Shift Is Periodic – Not because of tilt and orbit, but because of ripple tension resetting cycles.

Final Thought Test

If the Earth truly spun at such incredible speeds, and hurtled through space at 108,000 km/h around the Sun, And if the Solar System was moving through the galaxy at 828,000 km/h, And the galaxy was zooming through the universe at 2.1 million km/h— Why can a spider build a perfect web, in still air, on your balcony?

Let that ripple.

Appendix D: Quotes to Question Motion

A curated collection of reflections, paradoxes, and forgotten insights that gently shake the foundations of assumed motion.

History isn't short of brilliance—but it is often short of *courage* to question foundational beliefs. This appendix collects striking quotes—some famous, some overlooked, and a few reframed—to help the reader pause, reflect, and consider:

Is motion the truth—or just how we explain change?

Each quote is followed by a brief ripple commentary from the UED perspective.

1. Isaac Newton (1643–1727)

"I frame no hypotheses."

Newton's humility rings loud here. He constructed mathematical descriptions—not explanations. His refusal to hypothesize about gravity's cause left a gap science later filled with assumption, not discovery.

UED Ripple: What Newton left open, UED dares to resolve—not as hypothesis, but as consequence in stillness.

2. Albert Einstein (1879–1955)

"Time and space are modes by which we think, not conditions in which we live."

Though Einstein mathematized space-time, he confessed its nature as a cognitive mode—a lens of perception.

UED Ripple: Time is a threshold of ripple response. It doesn't *flow*—we phase through it. Motion is a perception, not a property.

3. Galileo Galilei (1564–1642)

"All truths are easy to understand once they are discovered; the point is to discover them."

Galileo helped lift observational thinking above theological obedience. Yet, his heliocentric ideas were still limited by the tools of the time.

UED Ripple: True discovery must go beyond motion to uncover what moves the perception of motion.

4. Niels Bohr (1885–1962)

"Everything we call real is made of things that cannot be regarded as real."

Quantum mechanics' founding father himself admitted that reality—at the deepest level—is not mechanical.

UED Ripple: The field doesn't move. It ripples. It resolves. Reality is emergent alignment, not solid or fixed.

5. Werner Heisenberg (1901–1976)

"What we observe is not nature itself, but nature exposed to our method of questioning."

Observation collapses potential into appearance. This isn't magic—this is ATT (Alignment Threshold Time).

UED Ripple: You don't see what *is*—you see what resolves into alignment with your observing field.

6. David Bohm (1917–1992)

"The notion that all these fragments are separately existent is evidently an illusion."

Bohm rejected particle-centric views. He emphasized wholeness and hidden variables.

UED Ripple: The universe is one structured stillness, not a fragmented dance of particles.

7. Nikola Tesla (1856–1943)

"If you want to find the secrets of the universe, think in terms of energy, frequency, and vibration."

Tesla's intuitive genius was never limited by academia.

UED Ripple: Motion is not translation. It is ripple participation. Every cause is vibrational tension seeking equilibrium.

8. Rupert Sheldrake (b. 1942)

"The 'constants' of nature may not be constant."

If constants aren't constant, is motion stable? Are orbits predictable? Or are we misreading field patterns as fixed trajectories?

UED Ripple: Constants are consequences, not absolutes. Ripple fields tune them—not the other way around.

9. Alan Watts (1915–1973)

"The only way to make sense out of change is to plunge into it, move with it, and join the dance."

But what if the dance is illusion? What if stillness was the music?

UED Ripple: True change isn't movement—it is resolution of field imbalance. Join the ripple, not the race.

10. Sandeep Chavan (b. 1980)

"The Earth doesn't move. When you move, you think it does."

This foundational statement inverts the default assumption and demands radical honesty in observation.

UED Ripple: You're not chasing motion. You're resolving misalignment in a still universe.

11. Anonymous Ancient Saying

"As above, so below."

Alignment. Reflection. Mirror principles. In ancient mysticism, this quote framed universal balance.

UED Ripple: Ripple fields echo this truth. The micro and macro don't move apart—they resonate across scales.

Final Ripple Thought

"Motion is not fundamental. Stillness is not passive. Stillness is not absence—it is origin."

Let these quotes ripple. Let them untangle the dogma. Let perception regain its dignity. Let motion be re-seen—not as truth, but as consequence.

Appendix E: References and Further Reading

For the courageous observer, curious skeptic, and ripple realist.

This book is a confrontation—not just with outdated physics but with the way we think about reality. Below is a set of resources, classics, radical perspectives, and supportive research that readers can explore to deepen their understanding or challenge the ripple view even further.

Primary UED Publications by the Author

These are foundational readings for the Universal Energy Dynamics (UED) framework and its ripple-centric reinterpretation of physical reality:

- The Moon: Co-author of Life's Script *Reframing lunar stillness and perception thresholds.*
- Chavan's Theory of Everything *Core framework of ripple dynamics, energy equilibrium, and structured vacuum fields.*
- Chavan vs Einstein: Ripple Beyond Relativity *Critique of relativistic assumptions with ripple-based alternatives.*
- Dark ≠ Hidden: Debunking Dark Matter through

Ripple Field Logic *A radical take on missing mass as unresolved ripple tension.*

- The Chavan Codex *Condensed insights from multiple works in ripple mechanics.*
- Post-Materialist Human Model *Applying ripple field theory to human consciousness and health.*
- The Human Machine & Human Fields *(Upcoming) Exploring ripple alignment in biology, emotion, and healing.*

Classical Scientific Sources (For Contrast)

These are important to understand the context from which UED diverges:

- Isaac Newton – *Philosophiæ Naturalis Principia Mathematica Foundation of classical mechanics and gravity.*
- Albert Einstein – *Relativity: The Special and General Theory Essential background to space-time concepts.*
- Niels Bohr & Werner Heisenberg – Quantum Mechanics Foundational Papers *Explore wave-particle duality, uncertainty, and observer dependence.*
- Stephen Hawking – *A Brief History of Time Popular cosmological narrative rooted in Big Bang and relativity.*

Alternative Thinkers & Radical Physics

- David Bohm – *Wholeness and the Implicate Order Quantum potential and the interconnected field model.*

- Rupert Sheldrake – *The Science Delusion* (aka *Science Set Free*) *Challenges dogmatic science and opens space for new models.*
- Thomas Kuhn – *The Structure of Scientific Revolutions How paradigms shift—and why resistance to new ideas is inevitable.*
- Immanuel Velikovsky – *Worlds in Collision Controversial rethinking of planetary history and dynamics.*
- Ervin Laszlo – *The Akashic Field: An Integral Theory of Everything Integrative vision resonant with ripple field thought.*

Philosophical & Metaphysical Explorations

- Alan Watts – *The Book: On the Taboo Against Knowing Who You Are Challenges the illusion of separateness in perception.*
- Lao Tzu – *Tao Te Ching The ancient dance of stillness and flow, rooted in field-like wisdom.*
- The Upanishads *Early Indian exploration of consciousness, stillness, and universal order.*
- J. Krishnamurti – *Freedom from the Known Tears apart conditioning, including inherited scientific thought.*

Experimental & Visual Tools

- Stellarium (Free Sky Mapping Software) *See the stars—real-time and historically. Reconfirm sky stillness.*

- Foucault Pendulum Installations *Visit museums; observe the device often used to "prove" Earth's spin—and question its interpretation.*
- Solar & Lunar Observations *Track sunrise/sunset data and moon phases over years—note field-like consistency vs claimed orbital drift.*

Your Own Mind – The Greatest Lab

"There is no deeper experiment than the one you are living." – *Sandeep Chavan*

In the end, ripple truth is not found in equations or telescopes alone, but in the resolution of perception, the stillness behind thought, and the courage to question what everyone else assumes.

Acknowledgments

This book is the ripple of many lives, minds, and moments. Though it may appear to bear a single name on its cover, the truth is far deeper—every idea here is a convergence of experience, questioning, and silent encouragement from those seen and unseen.

First and foremost, I bow to Stillness itself—the unshakable foundation beneath all illusions of motion. Without it, there would be no alignment, no observation, no book.

To my family, who often quietly observed my restless mind questioning the unquestioned—I know it wasn't always easy to understand, but your presence gave me the strength to keep writing.

To my students, past and present—you are my most honest audience. Your confusion, curiosity, and counter-questions shaped many of the arguments in this book. You reminded me that true teaching doesn't lie in answers, but in how deeply we explore a question.

To those who disagreed, debated, dismissed, or laughed—thank you. You were the necessary tension in the field. Without your resistance, my thoughts might never have reached the clarity they did.

To those who supported this work, encouraged it, edited drafts, or simply stayed up listening—I owe you more than citations. You are the silent heroes of this ripple.

To the independent thinkers across time—Newton, Einstein, Bohm, Tesla, Feynman, Krishnamurti, Buddha—you showed that even the greatest ideas are stepping stones, not destinations.

To the earth beneath my feet—thank you for not moving, even when the world insisted you did.

Finally, to the readers—thank you for picking up a book that dares to say "Maybe everything you believed... was motion in disguise." Whether you agree or not, your willingness to look again makes you part of the alignment. This book is yours now.

— *Sandeep Chavan*

About the Author

Sandeep Chavan is an independent physicist, educator, and prolific author known for challenging the foundations of modern science with clarity, courage, and deep philosophical insight. A mechanical engineer by training and a ripple observer by awakening, he is the originator of the Universal Energy Dynamics (UED) framework—a bold alternative to classical, relativistic, and quantum models.

With over 100 books and research reports to his name, Sandeep's work transcends disciplines, bridging science, consciousness, and ancient wisdom into a single coherent field theory. He speaks to thinkers, skeptics, students, and spiritual seekers alike—offering a refreshingly rational and radically intuitive understanding of the universe.

Beyond equations and experiments, his mission is simple yet audacious: Return the power of perception to the individual.

When not writing, Sandeep can be found teaching under a tree, sharing insights with curious minds, or quietly watching the stars that—he insists—haven't moved an inch.

Don't miss out!

Visit the website below and you can sign up to receive emails whenever Sandeep Chavan publishes a new book. There's no charge and no obligation.

https://books2read.com/r/B-A-EPGPC-ICMRG

BOOKS 2 READ

Connecting independent readers to independent writers.

Did you love *The Earth Doesn't Move*? Then you should read *Dark ≠ Hidden*[1] by Sandeep Chavan!

[2]

Dark ≠ Hidden*What if the universe wasn't withholding secrets—but we were just asking the wrong questions?*

For decades, physicists have been chasing shadows—"dark matter," "dark energy," and other invisible scaffolds—patching holes in models with concepts no telescope has ever truly confirmed. But what if the problem isn't what's missing... it's how we've been looking?

In *Dark ≠ Hidden*, **Sandeep Chavan**—engineer, philosopher, and creator of the **Universal Energy Dynamics**

1. https://books2read.com/u/m0Lq2Y
2. https://books2read.com/u/m0Lq2Y

(UED) framework—delivers a powerful alternative to the tired paradigms of modern physics. Rather than inventing unseeable particles or bending spacetime to fit data, Chavan introduces a field-aligned cosmos where every consequence is measurable, every ripple has memory, and nothing is truly dark—only misinterpreted.

Drawing from years of independent research and a bold critique of mainstream science, Chavan dismantles the myth of the "unknown" with precision and clarity. He offers fresh interpretations of:

Flat galaxy rotation curves—without invoking dark matter halosRedshift anomalies—as ripple fatigue, not cosmic accelerationGravitational lensing—explained by structured field memory, not exotic massVoid dynamics, cosmic flows, and high-z galaxies—through ripple alignment and entropy response

At the heart of this book lies a simple but revolutionary idea:

"You didn't fail to find it. You failed to ask what you were really seeing."

UED proposes a new physics rooted in **ripple response, Alignment Threshold Time (ATT),** and **Ripple Delay Index (RDI)**—measurable, testable, and observable tools that turn silence into signal and mystery into meaning. It's not a rejection of science. It's a return to what science always promised: **truth, without assumption**.

Written for researchers, thinkers, skeptics, and those exhausted by metaphysical placeholders, *Dark ≠ Hidden* is a declaration of intellectual freedom. It's an invitation to realign our instruments, our vocabulary, and most importantly—our curiosity.

Read more at https://www.sandeepjchavan.com/.

Also by Sandeep Chavan

Chavan's UED research

CMGED Research Report
Chavan's Cosmic Continuum (C^3)
Frequency
Ripples of Reality
Reality Realigned
The Great Realignment
I – The Cosmic Leverage
Chavan's Cosmic Theory of Relativity (CTR)
Chavan's Theory of Light
Feeling Like a Physicist
TIME – The New Definition
$E \neq mc^2$: The Collapse of Classical Assumptions
The Chavan Codex
Chavan's Atomic Model
Chavan's Theory of Everything

Echoes of War: The Russia-Ukraine Conflict and Its Global Ripples

Shattered Lives
The War Economy
Borders Redrawn
Arms and Allies
Collateral Damage
Justice in the Ashes

Somewhere in the Darkness of the Cosmos
The Blueprint of the Universe
The Cosmic Breath
From the Edge of the Universe

The Eternal Sage Chronicles
The Sage and the Sovereign
The Family Pillar
The People's Flame
The Warrior's Soul
The Bridging Mentor
The Guiding Light

The Exodus
The Last Cycle: 2075

The Fundamental Theory of Life & Existence

Energy in Action: The Soul Demystified
The Active Mind
Energy in Connection
Death: The Disconnection
Subconsciousness: The Chosen Disconnection
Desires: The Engine of Life

The Restoration of Human Knowledge
Restoring Bhagavad Gita

The War of Thoughts
Cracks in the Known
The Human Game
The Divine Dilemma
The Original Intent
The Answers

Ultra Modern Life Philosophy
The Duality of Laughter and Tears
Take It Easy
Compare to Accept

Standalone
It's Not AI, It's AHI - Amplified Human Intelligence

The Decision Paradox
The Four Sapiens
Malicious Script of Indian Polity
The IIT Legacy & Global Impact
The IIT Legacy & Global Impact
You are just a Version of Your Original
Win the Game You Didn't Choose
Always Keep Your Bags Packed
Reveal to Shield - Learn the Game of Facades
Engage Beyond Elections
Culture, Identity & Change: The Evolution of Indian Society
The Triangular Dynamics
Never Trust Without Doubt
If You Can't Change the Country, Leave the Country
Live Fearless with Fear
India: The Investment Magnet
Realignment
Powering a Billion Dreams
Universal Skill
Extension: From Brush to Algorithm
Human Side of Einstein
The Future Knows No Lines
Let's Bring the Generations to the Table
Unified Energy Dynamics
A Beat That Never Dies
Strategic Horizons
Right From Scratch
The Forgotten Innovators
Physics Fixing
The Savior Illusion
The Selfish Side

The Age of the Weakest Sapiens
Neural Wars
The Absorption: There Is No Death
Strange Therapist
Bubbles: Become a Drop
Beyond the Blue
Half-Scholar
The Law of Truth
Failure
Philosophy 2.0: A Structural Manifesto
Chaos: A Gentle Guide to the Unresolvable
Self-Help: New Syllabus
Rippling
Game: How to Play the Share Market
Out of Syllabus
Saffronized: The Code Behind the Crown
The Anger
Moksha: The Liberation
Chavan vs Einstein: The Universe Was Never Curved
Rocket Science: The Ripple Gambling
The Cosmos: The Still Universe
The Taste of Silence
Chavan's Post-Materialist Human Model
10-Second Rule
Chavan vs Hawking
It was Never a Wave Signal
Quantum Courtroom: The Trial of a Century
Chavan–Heisenberg Ripple Uncertainty Principle
The Great Disruption
Dark ≠ Hidden
Breath

Calculus of Life

The Earth Doesn't Move

Watch for more at https://www.sandeepjchavan.com/.

www.ingramcontent.com/pod-product-compliance
Lightning Source LLC
LaVergne TN
LVHW090354210926
848480LV00002B/341
9798231517015